SAMKHYA PHILOSOPHY

SAMKHYA PHILOSOPHY

Editor-in-Chief

Swami Chidatman Jee Maharaj

ANMOL PUBLICATIONS PVT. LTD.

NEW DELHI - 110 002 (INDIA)

ANMOL PUBLICATIONS PVT. LTD.
H.O.: 4374/4B, Ansari Road, Daryaganj,
New Delhi-110 002 (India)
Ph.: 23278000, 23261597
B.O.: No. 1015, Ist Main Road, BSK IIIrd Stage
IIIrd Phase, IIIrd Block,
Bangalore - 560 085 (India)
Visit us at: www.anmolpublications.com

Samkhya Philosophy

First Published, 2009
ISBN 978-81-261-3621-6

PRINTED IN INDIA

Printed at Mehra Offset Press, Delhi.

Contents

Preface

India colourful and vibrant, a land as diverse as its people. A mosaic of faiths, cultures, customs and languages that blend harmoniously to form a composite whole. One of the world's oldest living civilizations—which gave to the world—the concept of zero, the primordial sound Aum..., Yoga and Buddhism.

She is old. She is young. The two diametrically opposite, different yet unique and dynamic faces of India coexist even at this turn of the millennium. With a history of heritage, learning and civilisation dating back to almost 5000 years or perhaps even more, the modern India presents itself as an ideal country. From snow clad mountains to breezy sparkling blue sea side. From arid deserts with shifting sand dunes to lush green country sides. From ancient relics and architectural ruins dating back into antiquity to modern cities, India has it all.

To the north it is bordered by the world's highest mountain chain, where foothill valleys cover the northernmost of the country's maximum states. Further south, plateaus, tropical rain forests and sandy deserts are bordered by palm fringed beaches. Its southernmost tip is the meeting point of three Oceans.

India is a land of staggering contrast with a mingling of the tradition and modernity that is an unique experience to be in India. One will forever remember the stay in this wonderful Homeland-a land truly of mystery and enchantment. Only India can offer such an astonishing variety of contrasts.

She has given birth to numerous spiritual leaders and founders of some of the popular religions of the present day world. India probably has the most religious diversity in any country. It's the birthplace of Hinduism, Buddhism, Sikhism and Jainism and is among the few places to have a resident

Zoroastrian population. India with her veritable treasure trove of culture, mysticism, philosophy, art, music and architecture to name a few. Much of India's classical music is devotional and the North Indian Hindustani and South Indian Carnatic streams are distinct and both have a complex 'raga' framework. The legacy of dance in India is tremendous. The classical dances of India are numerous such as Kathakali of Kerala, Bharatnatyam of Tamil Nadu, Kuchipudi of Andhra Pradesh, Manipuri and Odissi from Orissa are the prominent dance forms in this country that sways to an altogether novel beat. The Yakshagana, nautanki and puppetry are ancient folk forms that live on till date. The earliest specimens of Indian painting are the ones on the walls of the Ajanta Caves dating back to 2nd century BC. The Mughals had a huge impact on Indian art. The influence of Persian art brought placid garden scenes, illustrations from myths, legends and history into Indian art. Word craft, handicrafts, architecture and sculpture all contribute to this rich and varied domain. Indian literature, both in English and in the vernacular, is ever more popular around the world.

The book aims at presenting perspectives of different aspects of Indian culture. It has topics on Indian classical music and dance, painting, ayurveda and yoga which are traditional and indigenous system of physical and mental wellbeing, language, philosophy and excellence achievements by Indian in the different branches of science and social sciences. It also includes our ancient religious and classic literature.

It is hoped that this book will be welcomed by the general readers as also scholars who are interested in the Indian heritage.

—Editors

1

Introduction

Samkhya, is one of the six schools of classical Indian philosophy. Sage Kapila is traditionally considered to be the founder of the Samkhya school, although no historical verification is possible. It is regarded as the oldest of the philosophical systems in India.

This was later incorporated as being one of the six orthodox (*astika*) (that which recognizes vedic authority) systems of Hindu philosophy with the major text of the theistic school being the extant *Samkhya Karika*, written by Ishvara Krishna, circa 200 CE. There are no purely Samkhya schools existing today in Hinduism, but its influence is felt in Yoga and Vedanta schools of philosophy. Its philosophy regards the universe as consisting of two eternal realities: Purusha and Prakriti. It is therefore a strongly dualist and enumerationist philosophy. The Purusha is the centre of consciousness, whereas the Prakriti is the source of all material existence.

The Samkhya school has deeply influenced the Hindu Yoga school of philosophy. They are sometimes referred together as Samkhya-yoga school.

Literature of Samkhya

Tradition ascribes the foundation of this school to the sage Kapila, but there is no evidence to prove the texts attributed to him, the *Samkhyapravacana Sutra* and the *Tattvasamasa* were actually composed by him. The earliest extant text of this school is *Samkhya Karika* of Ishvara (3rd century CE). Ishvara in his *Karika* described himself as being in the succession of the disciples from Kapila, through *Asuri and* Pancashikha. Gauapada wrote a commentary on this *Karika*. The next important work is Vacaspati's *Samkhyatattvakaumudi* (9th century CE). Narayana's treatise *Samkhyacandrika* is based on the *Karika*. The *Samkhyapravacana Sutra* is assigned to the 14th century CE as Gunaratna (14th century CE) did not refer this text but referred the *Karika*. This text has 6 chapters. Anirruddha's *Samkhyasutravtti* (15th century CE) and Mahadeva's *Samkhyasutravttisara* (c. 1600 CE) are the two important commentaries on the *Samkhyapravacana Sutra*. The most important commentary on the *Samkhyapravacana Sutra* is Vijnanabhikcu's *Samkhyapravacanabhacya* (16th century CE).

Evolution of the Samkhya School

Samkhya doctrines go back to older Upanishads, and were compiled into a formal system only later. The history of the evolution of this school of thought begins with dualist teachings in the Upanishads. Anima Sen Gupta characterizes the references to samkhya ideas in the Upanishads as "scattered", gradually being developed into an orthodox system.

A division by time periods is noted by the commentator Gunaratna (14th c. CE) who mention two Samkhya schools: *maulikya* (original) and *uttara* (late). Some Indologists use the terms "preclassical" and "classical" to distinguish between the early concepts and the codified system that eventually arose.

While many people believe that it was always an atheistic school of thought, in fact Samkhya passed through both theistic and atheistic stages of development as Gupta explains:

"In the classical Samkhya both dualism and atheism are visible in clear and vivid forms. The complete passivity and disinterestedness of Puruca and the acceptance of trigunatmika Prakti, as the independent cause of all inner and outer manifestations of the world, are the important characteristics of the classical form. It is also realistic in its attitude towards the phenomenal world. The pre-classical Samkhya on the other hand, has passed through different forms and stages such as theistic and monistic, atheistic and semidualistic, and so on."

There were several different schools of Samkhya, differing in their theistic emphasis and their conception of the soul.

Epistemology of Samkhya

According to the Samkhya school, all knowledge is possible through three *pramanas* (means of valid knowledge) (*Samkhya Karika*, *Sloka* 4):

1. Pratyaksha or Drishtam-direct sense perception,
2. Anumana-logical inference and
3. Sabda or Aptavacana-verbal testimony.

Samkhya cites out two types of perceptions, a. Indeterminate (*nirvikalpa*) perceptions and b. determinate (*savikalpa*) perceptions.

Indeterminate perceptions are merely impressions without understanding or knowledge. They reveal no knowledge of the form or the name of the object. There is only external awareness about an object. There is cognition of the object, but no discriminative recognition.

For example, a baby's initial experience is full of impression. There is a lot of data from sensory perception, but there is little or no understanding of the inputs. Hence they can neither be differentiated nor be labelled. Most of them are indeterminate perceptions.

Determinate perceptions are the mature state of perceptions which have been processed and differentiated appropriately. Once the sensations have been processed, categorized and interpreted properly, they become determinate perceptions. They can lead to identification and also generate knowledge.

Metaphysics of Samkhya

Ontology

Broadly, the Samkhya system classifies all objects as falling into one of the two categories: Purusha and Prakriti. Metaphysically, Samkhya maintains a radical duality between spirit/consciousness (*Purusha*) and matter (*Prakrti*).

- Purusha

Purusha is the Transcendental Self or Pure Consciousness. It is absolute, independent, free, imperceptible, unknowable, above any experience and beyond any words or explanation. It remains pure, "nonattributive consciousness". Purusha is neither produced nor does it produce.

• Prakriti

Prakriti is matter. Matter is inert, temporary, and unconscious. It is composed of three qualities (gunas) corresponding to creation, sustenance, and destruction. They are:

- sattva (goodness) – pure, elevating, enlightening
- rajas (passion) – motivates us to create, acquire and enjoy
- tamas (ignorance) – dirty, degrading, deluding, and destructive.

All physical events are considered to be manifestations of the evolution of *Prakriti*, or primal Nature (from which all physical bodies are derived). Each sentient being is a *Purusha*, and is limitless and unrestricted by its physical body. *Samsaara* or bondage arises when the *Purusha* does not have the discriminate knowledge and so is misled as to its own identity, confusing itself with the physical body-which is actually an evaluate of *Prakriti*. The spirit is liberated when the discriminate knowledge of the difference between conscious Purusha and unconscious Prakriti is realized.

• Ishwara (*Creationist God*)

The original school of Samkhya as founded by Sage Kapila. There has no philosophical place for a creationst God in this system. The Samkhyan's argue that the existence of Ishvara cannot be proved and hence cannot be admitted to exist. The school also argues that an unchanging Ishvara as the cause cannot be the source of a changing world as the effect.

Later on followers of Samkhya adopted theism and included Ishvara within the system. The concept of Ishvara was incorporated into the Samkhya viewpoint only after it became associated with the theistic Yoga system of philosophy.

Nature of Duality

According to Samkhya, the efficient cause of the world is Purusha and the material cause is Prakriti. Here Purusha stands for the 'Supreme Self' and Prakriti stands for 'Matter'. Purusha (Self) is the first principle of Samkhya. Prakriti is the second, the material principle of Samkhya.

Theory of Existence

The Samkhya system is based on Satkaryavada. According to Satkaryavada, the effect pre-exists in the cause. Cause and effect are seen as different temporal aspects of the same thing-the effect lies latent in the cause which in turn seeds the next effect.

More specifically, Samkhya system follows the Prakriti-Parinama Vada. *Parinama* denotes that the effect is a real transformation of the cause. The cause under consideration here is Prakriti or more precisely Mula-Prakriti (Primordial Matter). The Samkhya system is therefore an exponent of an evolutionary theory of matter beginning with primordial matter. In evolution, Prakriti is transformed and differentiated into multiplicity of objects. Evolution is followed by dissolution. In dissolution the physical existence, all the worldly objects mingle back into Prakriti, which now remains as the undifferentiated, primordial substance. This is how the cycles of evolution and dissolution follow each other.

Samkhya theorizes that Prakriti is the source of the world of becoming. It is pure potentiality that evolves itself successively into twenty four tattvas or principles. The evolution itself is possible because *Prakriti* is always in a state of tension among its constituent strands-

- *Sattva*-a template of balance or equilibrium;
- *Rajas*-a template of expansion or activity;
- *Tamas*-a template of inertia or resistance to action.

All macrocosmic and microcosmic creation uses these templates. The twenty four principles that evolve are-

- *Prakriti*-The most subtle potentiality that is behind whatever is created in the physical universe, also called "primordial Matter". It is also a state of equilibrium amongst the Three Gunas.

- *Mahat*-first product of evolution from Prakriti, pure potentiality. *Mahat* is also considered to be the principle responsible for the rise of *buddhi* or intelligence in living beings.
- *Ahamkara* or ego-sense-second product of evolution. It is responsible for the self-sense in living beings. It is also one's identification with the outer world and its content.
- "Panch Tanmatras" are a simultaneous product from Mahat Tattva, along with the Ahamkara. They are the subtle form of Panch Mahabhutas which result from grossification or Panchikaran of the Tanmatras. Each of these Tanmatras are made of all three Gunas.
- *Manas* or "Antahkaran" evolves from the total sum of the *sattva* aspect of *Panch Tanmatras* or the "Ahamkara"
- *Panch jnana indriyas* or five sense organs-also evolves from the *sattva* aspect of *Ahamkara*.
- *Panch karma indriya* or five organs of action-The organs of action are hands, legs, vocal apparatus, urino-genital organ and anus. They evolve from the *rajas* aspect of *Ahamkara*.
- *Panch mahabhuta* or five great substances-ether, air, fire, water and earth. They evolve from the "tamas" aspect of the "Ahamkara". This is the revealed aspect of the physical universe.

The evolution of primal Nature is also considered to be purposeful-Prakrti evolves *for* the spirit in bondage. The spirit who is always free is only a witness to the evolution, even though due to the absence of discriminate knowledge, he misidentifies himself with it.

The evolution obeys causality relationships, with primal Nature itself being the material cause of all physical creation. The cause and effect theory of Samkhya is called *Satkaarya-vaada* (theory of existent causes), and holds that nothing can really be created from or destroyed into nothingness-all evolution is simply the transformation of primal Nature from one form to another.

The evolution of matter occurs when the relative strengths of the attributes change. The evolution ceases when the spirit realizes that it is distinct from primal Nature and thus cannot evolve. This destroys the purpose of evolution, thus stopping Prakrti from evolving for Purusha.

Samkhyan cosmology describes how life emerges in the universe; the relationship between Purusha and Prakriti is crucial to Patanjali's yoga system. The evolution of forms at the basis of Samkhya is quite unique. The strands of Samkhyan thought can be traced back to the Vedic speculation of creation. It is also frequently mentioned in the Mahabharata and Yogavasishta.

Moksha

Like other major systems of Indian philosophy, Samkhya regards ignorance as the root cause of bondage and suffering (*Samsara.*) According to Samkhya, the Purusha is eternal, pure consciousness. Due to ignorance, it identifies itself with the physical body and its constituents-Manas, ahamkara and Mahat, which are products of Prakriti. Once it becomes free of this false identification and the material bonds, Moksha ensues.

Samkhya in the Srimad Bhagavatam

"In the Urimad-Bhagavatam, First Canto, it is described that one becomes joyful by discharging devotional service. In that joyful attitude, one can understand the science of God, or Krishna consciousness; otherwise it is not possible. The analytical study of the elements of material nature and the concentration of the mind upon the supersoul are the sum and substance of the Samkhya philosophical system. The perfection of this Samkhya-yoga culminates in devotional service unto the absolute truth." (Urimad-Bhagavatam 3.26.72 purport)

A.C. Bhaktivedanta Swami Prabhupada has given extensive commentary on the authoritative definition of Samkhya philosophy in his English translation of the Srimad Bhagavatam or Bhagavat Purana, which is considered the "ripened fruit of the tree of Vedic knowledge".

The path of Samkhya yoga has nothing to do with mental speculation and is not the creation of any conditioned living entity of this material world. Explanations of Samkhya philosophy devoid of the devotional service of the Supreme Lord or which refuse to acknowledge that Kapiladeva belongs to an entirely different class of entity (Vishnu tattva) cannot liberate anyone from material bondage. Such commentaries diverge fundamentally from the Vedic conclusion *ishvara parama krishna sac cid ananda vigraha*; a conclusion which clearly states that the highest truth is Lord Krishna, who is the cause of all causes and possesses an eternal body of perfect cognizance and bliss.

The real Samkhya philosophy is originally explained by Sri Kapiladeva, who is Lord Sri Krishna appearing Himself to distribute transcendental knowledge for the enlightenment of the conditioned souls. Lord Kapiladeva appeared as the son of Devahuti and Kadarma Muni, who were both elevated devotees. The mother of Kapiladeva, Devahuti, taking the role of a fallen soul, had expressed to her exalted son "My dear Kapila, I am very sick of the disturbance of my material senses, for because of this sense disturbance, my Lord, I have fallen into the abyss of ignorance. You have come as my son, but you are my guru because you can inform me how I can cross the ocean of nescience, which is the material world. Your Lordship is my only means of getting out of this darkest region of ignorance because You are my transcendental eye, which, by Your mercy only, I have attained after many many births. You are the Supreme Personality of Godhead, the origin and Supreme Lord of all living entities. You have arisen to disseminate the rays of the sun in order to dissapate the darkness of ignorance of the universe". After hearing his mother's uncontaminated desire for transcendental realization, the Lord thanked her within Himself for her questions, and thus, His face smiling, He explained the path of Samkhya-yoga for the transcendentalists, who are interested in self realization.." Urimad-Bhagavatam 3.25.8-14) Lord Kapiladeva goes on to explain that "The yoga system which relates to the Lord and the individual soul, which is meant for the ultimate benefit of the living entity, and which causes detachment from all happiness and distress in the material world, is the highest yoga system." (Urimad-Bhagavatam 3.25.13)

- On the highest platform of yoga one who becomes transcendentally situated by elevation to devotional service to the Lord is joyfully aloof from the temporary and ephemeral states of mundane happiness and distress. That state is described by Lord Krishna in the Bhagavad-Gita As It Is(18.54) *brahma-bhutah prasannatma na socati na kanksati samah sarvesu bhutesu mad-bhaktim labhate param* "One who is thus transcendentally situated at once realizes the Supreme Brahman. He never laments nor desires to have anything; he is equally disposed to every living entity. In that state he attains pure devotional service unto Me."

When the consciousness of the living entity is conditioned by the three modes of material nature, sattvas, rajas and tamas, the living entity is considered to be entangled in guneshu, or maya consciousness. When the same consciousness is attached to the Supreme Lord, one is considered to be in liberated consciousness. Material consciousness conditioned by the modes of nature has as its cental focus fruitive action by the living entity for the enjoyment of material sense gratification. Pure Krishna consciousness is the innate quality of the living entity and is characterised by the absence of fruitive desire for sense enjoyment, which is replaced by the desire to please the senses of the Supreme Lord.

The entire yogic process, culminating in bhakti yoga, is to cleanse the mind and senses of the effects of kama and lobha – lust and greed. As soon as the living entity is purified of the false ego of identifying with the material body he transcends the influence of mundane happiness and distress. The living entity is described in the Bhagavad Gita (15.7) as mamaivamso jiva loke jiva bhuta sanatana, the "eternal, fragmental part of the Supreme." Contrary to the monist school of thought, the living entity and Supreme Lord are eternally separate individuals. "The mayavada philosophy enunciates that the whole spirit exists, but a part of it, which is called the jiva, is entrapped by illusion. This philosophy is unnacceptable because spirit cannot be divided like a particle of matter. The part, the jiva, is eternally a part. As long as the Supreme Lord exists, His part and parcel also exists." (Urimad-Bhagavatam 3.25.17)

- Yogis and transcendentalists may follow numerous paths to attain jnana and vairagya (knowledge and renunciation), however, they cannot be successful without surrender and devotion to the Supreme Lord. (Srimad Bhagavatam 3.25.19)

Commentators who describe Kapiladeva as merely a sage or an unusually qualified person are disingenuous. Kapiladeva is the Supreme Personality of Godhead. Elsewhere on Wikipedia are unbonafide comments such as "The original school of Samkhya as founded by Sage Kapila. There is no philosophical place for a creationst God in this system. The Samkhyan's argue that the existence of Ishvara cannot be proved and hence cannot be admitted to exist. The school also argues that an unchanging Ishvara as the cause cannot be the source of a changing world as the effect." Such statements are not supported in the Urimad-Bhagavatam or the Bhagavad-Gita As It Is, both scriptures being considered the absolute essence of all Vedic wisdom.

It would be absurd for the Supreme Lord to propound a philosophical system which is diametrically opposed to the absolute nature of His existence as the creator and maintainer of the entire creation. Arguments otherwise misconstrue the Vedic conclusion and are entirely contrary to the authoritative Vedic scriptures such as the Urimad-Bhagavatam, which literally means, the beautiful story of the Supreme Lord. Complex word jugglery by mental speculators advertising themselves as great authorities on the Samkhya philosophy are useless. Descriptions of purusha which do not refer to the Supreme Purusha, Lord Krishna, are also useless. Descriptions of prakrti which do not refer to the fact that prakrti is the external energy of the Lord are also useless. The Bhagavad

Gita (7.4) clearly states that the material nature is the external, separated energy of the Lord, and that the living entities are part of His superior spiritual energy Bhagavad Gita (7.5) and that in both cases He is the original source.

Samkhya and Intelligent Design

Bhagavad-Gita and Urimad-Bhagavatam as translated by a bona fide spiritual master, A.C. Bhaktivedanta Swami Prabhupada, extensively describe intelligent design. The western tradition relies heavily on mental speculation to delve into matters which by definition lie beyond the purview of mundane logic.

Some proponents of intelligent design present the following argument about the fundamental nature of chance-that God must be intelligent enough to use statistical causation as a means of producing deterministic outcomes. This logic is the logic of God as the supreme clockmaker-he has created the universe, and then left it more or less to its own devices. In contrast to such an idea of a detached creator, the Vedic description, given by Sri Krishna Himself in Bhagavad-Gita, is that not only is He the supreme energetic source of all manifestation, but that he intimately controls the minutiae of His creation. In the Vedas this is called simultaneous oneness and difference. In essence, God is one with his creation, at the same time He retains His individual identity as the Supreme Being.

The material creation is described in Bhagavad-Gita As It Is as being composed of two types of God's energy. Firstly, the material elements of earth, water, fire, air, ether, mind, intelligence and false ego are considered the separated energy of God. Secondly, the living entities are described as God's apara prakrti, or superior, internal spiritual energy, which are qualitatively the same as the Lord, despite struggling with the inferior material nature. In any case, all these energies emanate from the supreme and are under his complete control, Lord Krishna clearly states, This material nature is working under my direction. (Gita 9.10)

Mental speculation, intellectual and logical research and so on may be useful tools to understand innumerable material topics, however they are useless in the comprehension of the absolute truth and the real nature of His creation and energies. Lord Krishna empowers His authorized representatives and disseminates the transcendental truth in bona fide revealed scripture which can only be understood by submissive reception of their enlightened message.

Endless argument and disputation between philosophers over the point of the existence or non existence of so called intelligent design is missing the real point-being that the absolute exists in complete independence of any puny mundane belief system or speculative intellectual process, and that the only real method of comprehension is to surrender to the Lord with faith and love, in a spirit of service. Lord Krishna describes in the Gita that an individual who adopts this approach will be enlightened by the Lord Himself-To those who are constantly devoted and worship me with love I give the understanding by which they can come to me. (Gita 10.10)

Evolutionary theory, various watered down creation arguments and other convoluted ideas, are all more or less expedient means of minimizing the incomprehensible power of the Lord. According to the Vedas, Krishna controls everything right down to the tiniest movements of the tiniest objects. Not a single thing happens by chance.

To acknowledge chance, the idea that God was somehow separated from his creation and therefore leaving events and outcomes to some type of statistical probability would have to be accepted. According to the Vedas, this is a misconception, as much as it is a misconception to reject reincarnation or the understanding that the soul in all living creatures is of the same type and nature.

Furthermore, mainstream theologians tend to be ephemeral at best on the question of the form and pastimes of God. The tendency to this type of formlessness or impersonalism is the antithesis of the Vedic paradigm, where the form, nature and pastimes of the Supreme Lord are clearly and extensively delineated. Instead of the form of God being described as a divine mystery of faith-(in other words, we don't know), the Srimad Bhagavatam and Brahma Samhita extensively describe the form and pastimes of the Lord, His abode and His associates.

The Vedas tell us that God has a name, form, qualities, pastimes and associates. He has everything we have, but to an infinitely greater and more sublime degree. After all, if everything we perceive in the mundane sphere is an emanation from the absolute truth, then clearly everything must also be possessed by the absolute truth. We have a form, pastimes and associates, yet we have the arrogance and audacity to deny the same of Krishna.

Finally, whether one accepts the Darwinian theory of evolution or not, common sense dictates that a Supreme intelligence is at work in the creation. An unlimited number of brilliant scientists, working without time restrictions with an unlimited budget, cannot create a single mosquito. To argue that the simplest of the Lord's creatures came about via random events or chance and that no superior intervention and design was involved is simply absurd.

The Vedas describe how the material creation is only a minor fraction of the entire creation of God – most of the creation is a variegated spiritual realm populated by the Lord, His expansions and His eternally liberated associates, who engage in spiritual pastimes of pure bliss and cognizance. Periodically the entire material creation is wound up into the body of Maha Vishnu, an expansion of Sri Krishna, the original source of all incarnations.

The intelligent creator of the manifested universe is neither Christian, Buddhist, Hindu or Moslem. The Supreme Personality of Godhead transcends all mundane sectarian considerations or categories.

The Samkhya Philosophy

Kapila was the founder of Samkhya philosophy. (8-6th Cent. B.C.E.?) Ishvara Krishna was its most famous writer. (3rd Cent. C.E.)

Basic Differences between Yoga and Samkhya

1. Samkhya is atheistic; Yoga is quasi-theistic Ishvara is simply the one *purusha* not to have fallen into *prakriti*.

2. Samkhya is a *jnana yoga*-route to salvation through knowledge only. Yoga is a *dhyana yoga*-meditation and asceticism. Patanjali's criticism of Samkhya: metaphysical knowledge alone will not save us.

Literal meaning of Samkhya is "discrimination" between *purusha* and *prakriti* so as to show the liberation of *purusha*. The focus is to discriminate between purusha and the higher mental states: *manas* and *buddhi*, which are part of *prakriti*. *Prakriti* is a fully real material substance, and not the creation of Brahman's uncanny power. Metaphysical "realism," *i.e.*, the external world is real. Metaphysical pluralism, too—*i.e.*, there are many individual souls that will remain individual and isolated even after their liberation from *prakriti*. Unlike Advaita Vedanta, pure *purusha* selves are indissolvably many.

At the liberation of the last *purusha* from *prakriti*, *prakriti* will return to its primoridal state. Its manifold appearance depends on our ignorance that we basically belong to it. But with proper knowledge and discrimination one can use *prakriti* for one's liberation.

Purusha has no attributes except that "it is" and that "it knows." "The spirit is what is sees, it is isolated, indifferent, a mere inactive spectator." (Quoted in Eliade, 27) It has no intelligence (this is located in *buddhi*.) and it is without desire. It is pure freedom. How then did it get enslaved? Originally, the three *gunas* (*sattva, rajas, tamas*) are in perfect equilibrium in *prakriti*. But under *purusha*'s influence disequilibrium and evolution begin.

1. Metaphor of the lame woman (*prakriti*) being carried by the blind man (purusha).

2. Spectator (*purusha*) entranced by the dancer (*prakriti*).

Prakriti is the source of the world of "becoming" (change and motion) and sensation. *Purusha* represents true being: inactive, unchanging, and pure.

Sattvaguna is the "stuff" of consciousness and all higher mental states (associated with Hindu god Vishnu and the goddess Lakshmi); *rajasguna* is the source of activity, sensation, and emotion (associated with Brahma and the goddess Saraswati); and *tamasguna* is the source of resistance, inertia, and dissolution (associated with Shiva and the goddess Kali).

The Samkhya-Karika by Ishvara Krishna

(*Source Book*, Chap. 12)

The main text comes from the Third Century C.E. and the commentary in smaller type comes from about 850 C.E.

I. Three types of pain: (1) pains of body and mind; (2) pains caused by external nature; and (3) supernatural agents.

II. Vedic religious rites are inefficient in removing the three types of pain. Impurities arise because of the killing of animals and seeds. Samkhya is obviously pushing the doctrine of *ahimsa* (non-injury). On this point, Samkhya is contrary to the Veda, "and hence is better." How did this ever become an orthodox system?

III. *Prakriti* is a cause but not an effect. 7 causes and effects, 16 effects only; 25 total metaphysical elements. *Purusha* is neither cause nor effect.

Prakriti cannot have a cause or there would be an infinite regress, which was irrational for all ancient philosophers. See Aristotle-Aquinas in West.

IV. Samkhya Theory of Knowledge (Epistemology). Three means of correct knowledge: perception, inference (Samkhya syllogism), and valid testimony.

VI. Knowledge beyond the senses comes from inference based on analogy.

VIII. *Prakriti* cannot be perceived, not because of its non-existence, but because of its subtlety. It is perceived in its effects.

XI. *Gunas* aren't really qualities or attributes, they are constituents. *Prakriti* is a three-strand rope.

XIII. *Sattva*-buoyant and illuminating; *rajas*-exciting and mobile: *tamas*-sluggish and enveloping.

Analogy of wick, oil and light-opposing things do work together. *Prakriti* is only a means to *purusha's* en-lightenment.

The Arguments for *Purusha*

Its Necessary Existence and its Nature

(*Samkhya-karika* xvii)

I. All composite things are for another's use.

II. There must be absence of the three gunas.

III. Because there must be control.

IV. Because there must be someone to experience.

V. Because there must be a tendency to isolation. Appeal to scripture here.

The Arguments in Deductive Form

I. 1. All composite objects are for another's use.

2. All of nature is composite (*i.e.*, made of *gunas*).

3. If the user in #1 is composite, there would be an infinite regress of composite users.

4. A simple (non-composite) thing must exist apart from Nature.

One does not appear to be true unless we interpret, as Samkhya probably does, all of nature to be a means to the soul's end. "From Brahma to a blade of grass, the creation is for the soul's benefit, until one attains to the supreme knowledge" (*Samkhya-pravacana-Sutra*, quoted in Eliade, p. 20).

Plato's argument for the immortality of the soul:

1. A thing can be destroyed only by separating its parts.

2. The soul has no parts.

3. The soul can not be destroyed.

II. Being non-composite logically implies that spirit (*purusha*) has, referring back to section xi:

1. no attributes (*gunas*), because they would be "parts," especially in the image of the three-strand rope.

2. it's not "productive"-only prakriti produces; only a composite thing can produce.

3. If it is not productive, it's not active.

4. It's non-objective (*i.e.*, subjective), always a subject, never an object.

5. It's distinguishable. The principle of the identity of indiscernables:If two things have identical properties, then they must be identical.

6. It's uncommon (specific).

7. It's sentient-pure awareness.

7 seems OK if sentience means "pure awareness," but not OK if we take the etymological meaning as "capable of sensing or perceiving." Is this sensible perception: "She [*prakriti*] has been seen by me (*purusha*) (lxvi).

III. The argument from control

1. Nature does not control herself.

2. But there is control, *e.g.* we try to stop pain.

3. A controller independent from nature must exist.

Isn't the *jiva* self responsible for controlling. This would make the soul active, which it presumably is not.

IV. The argument from experience to experiencer.

1. We have experiences of pain and pleasure.

2. There must be something to experience pain and pleasure.

Question: Doesn't the *jiva* self do this? Parable of the two birds: the enjoyer and the one who simply looks on.

V. Argument from ascetic isolation. Valid testimony of scripture and seers. There is something beyond nature, *viz.*, an isolated spirit.

XVIII. *Purashas* must be plural because of (1) different times of birth and death; (2) different bodies engaging in action; (3) different proportion of *gunas*. Only proves plurality of *jiva* souls?

XIX. Purusha is isolated, neutral, seer, (spectator), inactive.

Contradicts most of XVII? Experiences no pleasure, pain or delusion.

XXII. Emanation out of *Prakriti*

Mahat (The Great Principle)

Ahamkara ("I"-principle) Set of 16—11 sense organs and 5 primary elements.

XXIII. *Sattva*-virtue, wisdom, nonattachment possession of lordly poweres (goodness). *Tamas* is the reverse.

XXXIX & XL Doctrine of Subtle body. The spritual body is invisible "formed primevally, (1) unconfined, (2) lasting, composed of will, (3) migrates, is devoid of experience's, and is invested in dispositions.

1. Subtle bodies for *purushas* were formed in the first emanations of *prakriti.*

2. The subtle body can go through anything.

3. If not willful and have dispositons, it would not migrate.

How can something with dispositions not have experiences?

Does the subtle body disappear in final liberation? Yes, because *prakriti* contracts completely at the end. It does "merge" back into nature.

LIII. Celestial evolution has 8 forms (*sattva*) (levels) animal (*tamas*) evolution 95) and one human form (*rajas*).

LVI. Evolution is done by *prakriti* itself. Not by God, not Brahman, and not by Ishvara the Lord.

LVI. *Prakriti* is like a cook who retires after making the meal. Nature, properly understood, provides "nourishment" for *purusha* on its way to liberation. The *purusha* who realizes that is needs no more sustenance will then be free from nature, putting the "cook" out of work.

LVII. If God is a creator, he would create only happy mortals. And if mortals were in pain, he would be obligated to eliminate it. Thus, God cannot be involved in the operation of *prakriti*. As *prakriti* is basically insentient, it cannot be blamed for pain and evil. Similar solution to the problem of evil in Aristotle and contempoary "process" theology.

LIX. Dancer metaphor

LXI. *Prakriti* as bashful. "I have been seen" and never appears to purusha again.

LXII. Strictly speaking, [King analogy] no *purusha* is ever bound or liberated or migrates. Illusion?

LXIV. Paradoxical Results of Samkhya analysis.

"I do not exist, naught is mine, I am not." (1) Nothing is known (outside of *purusha*). (2) But I am pure, free from ignorance, and I am absolute.

1. I do not exist as *prakriti* exists. I am not matter, nature, or even high mental states.

2. Only the elements of *prakriti* can be known, and *purusha* is not part of *prakriti*, and finally has no relation with it.

LXVI. "She has been seen by me," and I [*purusha*] have lost interest. "I have been seen," says *prakriti* and she stops her dance (evolution).

The Samkhya

Obeisance to Sri Kapila Muni, the founder of the Samkhya system of philosophy, the son of Brahma, the Avatara of Vishnu.

The word 'Samkhya' means 'number'. The system gives an enumeration of the principles of the universe, twenty-five in number. Hence the name is quite appropriate. The term 'Samkhya' is used in the sense of 'Vichara' or 'philosophical reflection' also.

In the Samkhya system, there is no analytical enquiry into the universe as actually existing, arranged under topics and categories. There is a synthetical system, starting from an original primordial Tattva or Principle, called Prakriti, that which evolves or produces or brings forth (Prakaroti) everything else.

Perception (Pratyaksha), inference (Anumana) and right affirmation (Apta Vakya) are the three Pramanas or proofs in the Samkhya system. The word Apta means fit or right. It is applied to the Vedas or inspired teachers. The Naiyayikas have four kinds of proofs, *viz.*, perception, inference, comparison and verbal authority. The Mimamsakas recognise six kinds of proofs.

Dual Concept of Purusha and Prakriti

The Samkhya system is generally studied next to the Nyaya. It is a beautiful system of philosophy. The Western philosophers also have great admiration for this system. It is more categorically dualistic. It denies that anything can be produced out of nothing. It assumes the reality of Purusha and Prakriti, the knowing Self and the objects known.

Prakriti and Purusha are Anadi (beginningless) and Ananta (infinite). Non-discrimination between the two is the cause for birth and death. Discrimination between Prakriti and Purusha gives Mukti (salvation). Both Purusha and Prakriti are Sat (real). Purusha is Asanga (unattached). He is consciousness, all-pervading and eternal. Prakriti is doer and enjoyer. Souls are countless.

Non-acceptance of Isvara or God

The Samkhya system is called Nir-Isvara (God-less) Samkhya. It is atheistic. The Sankhyas do not believe in Isvara. They do not accept Isvara (God). The creation produced by Prakriti has an existence of its own, independent of all connection with the particular Purusha to which it is united. So the Sankhyas say that there is no need for an intelligent Creator of the world, or even of any superintending power.

This is a mistake; according to the Vedanta, Prakriti is always under the control of the Lord. It cannot do anything by itself. The Lord gazes at Prakriti. Then alone it is put in motion, and it begins to create. Prakriti is non-intelligent. An intelligent Creator alone can have a thought-out plan for the universe. Prakriti is only a helper (Sahakari). This is the theory of Vedanta.

Theory of Evolution and Involution

The Samkhya adopts the theory of evolution and involution. The cause and effect are the undeveloped and developed states of one and the same substance. There is no such thing as total destruction. In destruction, the effect is involved into its cause. That is all.

There cannot be any production of something out of nothing. That which is not cannot be developed into that which is. The production of what does not already exist potentially is impossible like a horn on a man, because there must, of necessity, be a material out of which a product is developed, and because everything cannot occur everywhere at all times, and also because anything possible must be produced from something competent to produce it.

That which does not exist cannot be brought into existence by an agent. It would be useless to grind groundnut, unless the oil existed in it. The same force applied to sand or orange would not express groundnut oil. The manifestation of the oil is a proof that it was contained in the groundnut and consequently, a proof of the existence of the source from which it is derived.

The effect truly exists beforehand in its cause. This is one of the central features of the Samkhya system of philosophy. Cause is a substance in which the effect subsists in a latent form. Just as the whole tree exists in a latent or dormant state in the seed, so also the whole world exists in a latent state in Prakriti, the Avyakta (unevolved), or the Avyakrita (undifferentiated). The effect is of the same nature as the cause. The effect or the product is not different from the material of which it is composed.

Fourfold Classification of the Twenty-five Tattvas (principles)

The Samkhya gives a description of categories based on their respective productive efficiency, *viz.*,

1. Productive (Prakriti)
2. Productive and Produced (Prakriti-Vikriti)
3. Produced (Vikriti)
4. Neither Productive nor Produced (Anubhayarupa)

This fourfold classification includes all the twenty-five principles or Tattvas. Prakriti or Nature or Pradhana (chief) is purely productive. It is the root of all. It is not a product.

It is a creative force, evolver, producer. Seven principles-intellect (Buddhi), egoism (Ahankara) and the five Tanmatras (subtle rudimentary elements in an undifferentiated state before quintuplication or Panchikarana)-are productions and productive. Buddhi (intellect) is productive as Ahankara (egoism) is evolved out of it. It is produced also, as it itself is evolved out of Prakriti. Egoism is a production, as it is derived from intellect.

It is productive, as it gives origin to the five subtle rudiments or Tanmatras. The subtle rudiments are derived from egoism. Hence they are productions. They give origin to the five elements (space, fire, air, water, earth). Hence they are productive. The sixteen principles, the ten organs, the mind and the five elements, are productions only. They are unproductive, because none of them can give birth to a substance essentially different from itself. The Purusha or Spirit is neither a production, nor is it productive. It is without attributes.

The Object of the Samkhya Philosophy

The enquiry into this system of philosophy is to find out the means for eradicating the three sorts of pain, *viz.*, internal or Adhyatmika (*e.g.*, fever and other diseases), celestial or Adhidaivika (lightening, cold, heat, floods, earthquakes, storms etc.), and external or Adhibhautika (pain from animals, snakes, scorpions etc.), and the disease of rebirths. Pain is an embarrassment. It stands in the way of doing Yoga Sadhana and attaining Moksha or release. Kapila Muni imparted a knowledge of the twenty-five principles which annihilated this pain. According to the Samkhya, he who knows the twenty-five principles attains liberation. The ultimate cessation of the three kinds of pain is the final goal of life.

Prakriti

'Prakriti' means that which is primary, that which precedes what is made. It comes from 'Pra' (before) and 'Kri' (to make). It resembles the Vedantic Maya. It is the one root of the universe. It is called Pradhana or the chief, because all effects are founded on it and it is the root of the universe and of all objects.

Characteristics of Prakriti

Pradhana or Prakriti is eternal, all-pervading, immovable. It is one. It has no cause, but is the cause of all effects. Prakriti is independent and uncaused, while the products are caused and dependent. Prakriti depends only on the activity of its own constituent Gunas (metaphysical properties).

Prakriti is destitute of intelligence. It is like a string of three strands. The three Gunas form the three strands. Prakriti is mere dead matter which is equipped with certain potentialities due to the Gunas.

The Modifications of Prakriti

Crude matter is without form. Mahat or the Cosmic Intelligence is its first form. Intellect is the matter for egoism. Egoism is a form of intellect. It is the matter of which the senses and the rudimental elements are formed. The senses and the rudimental elements are forms of egoism. The gross elements are forms of the rudimental elements. Intellect, egoism and the five subtle rudiments or Tanmatras are the effects of Prakriti. This creation, from intellect down to the elements, is brought about by the modifications of Prakriti. Having observed the effects, the cause (Prakriti) is inferred. It is imperceptible from its subtlety. It must, therefore, be inferred from its effect.

The Function of Prakriti

Prakriti is the basis of all objective existence. Prakriti does not create for itself. All objects are for the enjoyment of the spirit or soul. Prakriti creates only when it comes into union with Purusha, like a crystal vase with a flower. This work is done for the emancipation of each soul. As it is the function of milk to nourish the calf, so it is the function of Prakriti to liberate the soul.

The Gunas

According to the Samkhya philosophy, Prakriti is composed of three Gunas or forces, called Sattva (purity, light, harmony), Rajas (passion, activity, motion), and Tamas (inertia, darkness, inertness, inactivity).

Guna means a cord. The Gunas bind the soul with a triple bond. These Gunas are not the Nyaya-Vaiseshika Gunas. They are the actual substances or ingredients, of which Prakriti is constituted. They make up the whole world evolved out of Prakriti. They are not conjoined in equal quantities, but in varying proportions, one or the other being in excess. Just as Sat-Chit-Ananda is the Vedantic trinity, so also the Gunas are the Samkhya trinity.

Interaction Between the Gunas Leads to Evolution

The three Gunas are never separate. They support one another. They intermingle with one another. They are intimately related as the flame, the oil and the wick of a lamp. They form the very substance of Prakriti. All objects are composed of the three Gunas. The Gunas act on one another. Then there is evolution or manifestation. Destruction is only non-manifestation.

The Gunas are objects. Purusha is the witness-subject. Prakriti evolves under the influence of Purusha. Mahat or the great (Intellect), the Cause of the whole world, is the first product of the evolution of Prakriti. Ahankara (egoism) arises after Buddhi. Agency belongs to Ahankara. It is the principle that creates individuality. Mind is born of Ahankara. It carries out the orders of the will through the organs of action (Karma Indriyas). It reflects and doubts (Sankalpa-Vikalpa). It synthesizes the sense data into precepts. The mind takes part in both perception and action. There is no separate Prana Tattva in the Samkhya system. The Vedanta system has a separate Prana Tattva. In the Samkhya system, mind, with the organs, produces the five vital airs. Prana is a modification of the senses. It does not subsist in their absence.

Characteristics of the Three Gunas

Sattva is equilibrium. When Sattva prevails, there is peace and tranquillity. Rajas is activity which is expressed as Raga-Dvesha, likes or dislikes, love or hatred, attraction or repulsion. Tamas is that binding force with a tendency to lethargy, sloth and foolish actions. It causes delusion or non-discrimination.

When Sattva is predominant, it overpowers Rajas and Tamas. When Rajas is dominant, it overpowers Sattva and Tamas. When Tamas is predominant, it overpowers Rajas and Sattva.

How Man is Affected by the Three Gunas

There are three Gunas in every man. Sometimes, Sattva prevails in him. Then he is calm and tranquil. He reflects and meditates. At other times, Rajas prevails in him and he does various sorts of worldly activities. He is passionate and active. Sometimes, Tamas prevails. He becomes lazy, dull, inactive and careless. Tamas generates delusion. Again, one of these Gunas is generally predominant in different men. A Sattvic man is virtuous. He leads a pure and pious life. A Rajasic man is passionate and active. A Tamasic man is dull and inactive. Sattva makes a man divine and noble. Rajas makes him thoroughly human and selfish, and Tamas makes him bestial and ignorant. There is much Sattva in a sage or saint and there is much Rajas in a soldier, politician and businessman.

The Purusha

Characteristics of the Purusha

The Purusha or the Self is beyond Prakriti. It is eternally separate from the latter. Purusha is without beginning or end. It is without attributes and without qualities. It is subtle and omnipresent. It is beyond mind, intellect and the senses. It is beyond time, space and causality. It is the eternal seer. It is perfect and immutable. It is pure consciousness (Chidrupa).

The Purusha is not the doer. It is the witness. The Purusha is like a crystal without any colour. It appears to be coloured by the different colours which are placed before it. It is not material. It is not a result of combination. Hence it is immortal. The Purushas or souls are infinite in number, according to the Samkhya. There are many Purushas. If the Purushas were one, all should become free if any one attained release.

The different souls are fundamentally identical in nature. There is no movement for the Purusha. It does not go anywhere when it attains freedom or release. Souls exist eternally separate from each other and from Prakriti. Each soul retains its individuality. It remains unchanged through all transmigrations. Each soul is a witness of the act of a separate creation, without taking part in the act. It is a looker-on uniting itself with the unintelligent Prakriti, like a lame man mounted on a blind man's shoulders, in order to behold the phenomena of a creation, which Prakriti herself is not able to observe.

The Purusha or the Self is the witness (Sakshi), a spectator (Drashta), a by-stander (Madhyastha), solitary (Kaivalya), passive and indifferent (Udasina).

Inference of the Existence of the Purusha

Intelligence cannot belong to the intellect, because the intellect is material and is the effect of Prakriti which is non-intelligent. If intelligence is absent in the cause, it cannot manifest itself in the effect. Therefore, there must be a distinct principle of intelligence and this principle is Purusha or the Self.

The insentient body seems sentient on account of its union with the Self, and the Self appears as the agent. Just as a pot with cold water appears to be cold, with hot water seems to be hot, so intellect and the rest seem to be sentient on account of union with the Purusha. This mutual transfer of properties is like that of fire and iron, or that of the sun and water.

There must be a Supervisor over and above the Pradhana or Prakriti. The Supervisor is Purusha or the Self. Prakriti and its products are objects of enjoyment. There must exist an enjoyer who must be an intelligent principle. This intelligent enjoyer is Purusha or the Self.

Just as chair and bench are for the use of another, so also this body, senses and mind are for the use of the Self which is immaterial, as it is destitute of attributes and as it is beyond the Gunas. The Purusha is the witness of the Gunas. The Gunas are the objects. Purusha is the witness-subject. Hence, it is not affected by pleasure, pain and delusion which are attributes of the three Gunas, Sattva, Rajas and Tamas, respectively. If pain is natural to the Purusha and if the Purusha is not naturally free from the action of the Gunas, no salvation from rebirth is possible.

Purusha and Prakriti – A Contrast

The characteristics of Prakriti and Purusha are contrary in nature. Purusha is consciousness, while Prakriti is non-consciousness. Purusha is inactive (Akarta), while Prakriti is active. Purusha is destitute of the Gunas, while Prakriti is characterised by the three Gunas, Purusha is unchanging, while Prakriti is changing. The knower is Purusha. The known is Prakriti. The knower is the subject or the silent witness. The known is the visible object.

The Universe

The world is evolved with its different elements when the equilibrium in Prakriti is disturbed. The countless Purushas exert on Prakriti a mechanical force which distracts the equipoise of Prakriti and produces a movement. Then the evolution of the universe starts.

The Process of Evolution and Involution

Prakriti is the root of the universe. Prakriti is both the material and the efficient cause of the universe. From this Prakriti emanates the cosmic Buddhi or Mahat. From Mahat proceeds the cosmic Ahankara or the principle of egoism. From this egoism emanates the ten senses and the mind on the subjective side, and the five subtle Tanmatras of sound, smell, taste, vision (or colour) and touch on the objective side. From these Tanmatras proceed the five gross elements-earth, water, fire, air and space.

Akasha (space) has the property of sound which is the Vishaya or object for the ear. Vayu (air) has the property of touch which is the Vishaya for the skin. Tejasa (fire) has the property of form or colour which is the Vishaya for the eye. Apas (water) has the property of taste which is the Vishaya for the tongue. Prithvi (earth) has the property of odour which is the Vishaya for the nose. Each of these elements, after the first, has also the property of preceding besides its own.

During dissolution of the world, the products return by a reverse movement into the preceding stages of development, and ultimately into Prakriti. Earth merges in its cause, water, water in fire, fire in air, air in space; and space in Antahkarana (egoism), and Antahkarana in Mahat (intellect), and Mahat in Prakriti This is the process of involution. There is no end to Samsara or the play of Prakriti. This cycle of evolution and involution has neither a beginning nor an end.

The Process of Knowledge

An object excites the senses. The mind arranges the sense impressions into a percept. Egoism refers it to the Self. Intellect forms the concept. It converts the precept into a concept and presents it to the Purusha. Then there is knowledge of the object.

Before you engage in any matter, you first observe or consider, then you reflect, and then determine: "This must be done by me"; and then you proceed to act. This ascertainment: "Such act is to be done by Me", is the determination of the intellect (adhyavasaya). The intellect is an instrument which receives the ideas or images conveyed through the organs of sense and the mind, constructs them into a conclusive idea, and presents this idea to the Self. The function of the intellect is determination (Nischaya).

The mind is both an organ of sensation and of action. The senses receive simple impressions from without. The mind cooperates with the senses, and then the impressions are perceived. The mind ponders, the intellect determines, and egoism becomes conscious.

Agency belongs to egoism-the Ahankara or the I-maker-which is itself a product of Prakriti, but not to the Purusha or Self who is always a silent witness.

Intellect, egoism, mind and the eye see a form (shape) at once, in one instant, and come immediately to the conclusion, say, "This is a jar". The same three, with the tongue, at once relish taste; with the nose smell; and so with the ear and the skin. The function is also occasionally gradual. A man going along a road sees an object at a distance. A doubt arises in his mind whether it is a wooden pole (post or a stump) or a man. He then sees a bird sitting on it. Then the doubt is removed by the reflection of the mind.

The intellect makes a determination that it is a wooden pole only. Then the egoism says: "I am certain it is a wooden pole only". In this way, the functions of the mind, intellect, egoism and the eye are gradual, also. There is leisure for the eye to see, for the mind to reflect or consider, for egoism to apply, and for the intellect to conclude. There is another example. The ear hears the twang of a bowstring; the mind reflects that this must be for the shooting of an arrow; egoism says: "It is aimed at me"; and the intellect determines: "I must run at once".

The intellect, the mind and egoism are the door-keepers. The five senses of perception or Jnana-Indriyas are the gates. The intellect is the instrument or organ which is the medium between the senses and the Self.

The Intellect and its Functions

The intellect or the Buddhi is the most important of all the products of Prakriti. The senses present their objects to the intellect. The intellect exhibits them to the Purusha. The intellect discriminates the difference between Purusha and Prakriti.

The intellect is the instrument or organ which is the medium between the other organs and the Self. All ideas derived from sensation, reflection, or consciousness are deposited in the chief or great instrument, intellect, before they can be made known to the Self for whose use and advantage alone they have assembled. They convey impressions or ideas with the properties or effects of pleasure, pain and indifference, accordingly as they are influenced by the qualities of Sattva (purity), Rajas (passion) or Tamas (darkness).

Just as the headman of a village collects the taxes from the villagers and pays them to the collector of the district, just as the local collector pays the amount to the minister, and the minister receives it for the use of the state, so also the mind receives the ideas from the external organs, transfers them to egoism, delivers them to the intellect which is the general superintendent and takes charge of them for the use of the Sovereign Self. The intellect is the prime minister of Purusha. It brings for Purusha the fruition of all that is to be experienced. It appears to be intelligent on account of the reflection of Purusha which is very near to it, though, by itself, it is really non-intelligent.

The Jiva (embodied soul)

The Jiva is the soul in union with the senses. It is limited by the body. It is endowed with egoism. The reflection of Purusha in the Buddhi or intellect appears as the ego or the empirical soul. It is associated with ignorance and Karma. It is subject to pleasure and pain, action and its fruits, and rotates in the cycle of births and deaths.

The Jiva must realise the perfection of the Purusha. It must attain to the status of the Purusha. Every Jiva has in it the higher Purusha hidden within. It must become conscious of the real nature of the higher Purusha. Freedom or perfection is a return into one's true Self. It is the removal of an illusion which conceals one's true nature.

Release

Bondage belongs to Prakriti, but is attributed to Purusha. Purusha is eternally free. Union of Purusha with Prakriti due to non-discrimination is bondage; the failure to discriminate between Purusha and Prakriti is the cause of Samsara or bondage; and disunion of Purusha and Prakriti due to discrimination is emancipation. Release is not merging in the Absolute, but isolation from Prakriti.

The object of the Samkhya system is to effect the liberation of the Purusha or Self from the fetters which bind it on account of its union with Prakriti. This is done by conveying the correct knowledge of the twenty-four constituent principles f creation, and tightly discriminating the Self from them.

In the Samkhya system, the Pramanas or means of obtaining the correct measure of existing things, are three, *viz.*, Pratyaksha or perception by the senses, Anumana (inference) and Apta-Vachana (trustworthy testimony).

How Release is Effected?

When the separation of the soul from the body takes place by destruction of the effects of virtue, vice and the rest, and Prakriti ceases to act in respect to it, then there is the final and absolute emancipation or the final beatitude.

When the fruits of acts cease, and body-both gross and subtle-dissolves, Nature does not exist with respect to the individual soul. The soul attains the state called Kaivalya. It is freed from the three kinds of pain.

The Linga-Deha or subtle body which migrates from one gross body to another in successive births, is composed of intellect, egoism, mind, the five organs of knowledge, the five organs of action and the five Tanmatras. The impressions of actions done in various births are imbedded in the subtle body. The conjunction of the Linga-Deha with the gross physical body constitutes birth and separation of the Linga-Deha from the gross physical body is death.

This Linga-Deha is destroyed by the knowledge of the Purusha.

When one attains perfect knowledge, virtue and vice become destitute of causal energy, but the body continues for some time on account of the previous impulse, just as after the action of the potter has stopped, the wheel continues to revolve owing to the momentum given to it.

Release Is Nothing but Termination of the Play of Prakriti

The union of the Self with Nature or Prakriti is like the association of a lame man with a blind man. A lame man and a blind man were deserted by their fellow-travellers in a forest. They agreed to divide between them the duties of walking and of seeing. The lame man mounted himself on the shoulders of the blind man and directed the blind man. The blind man was able to pursue his route by the directions of his friend. Even so, the Self is like the lame man. The faculty of seeing is in the Self, not that of moving. The faculty of moving, but not of seeing, is in Prakriti. Prakriti is like the blind man. The lame man and the blind man separated when they reached their destination. Even so, Prakriti, having effected the liberation of the Self, ceases to act. The Self obtains Kaivalya or the final beatitude. Consequently, their respective purposes being effected, the connection between them terminates. The Self attains liberation by knowledge of Prakriti.

Prakriti's performances are solely for the benefit and enjoyment of the Self. Prakriti takes hold of the hand of the Self and shows it the whole show of the universe, and makes it enjoy everything which this world can give, and lastly helps it in its liberation.

In truth, the Self is neither bound nor released, nor does it migrate, but Nature alone in relation to various beings is bound, is released, and migrates.

As a dancing girl, having exhibited herself to the spectators, stops dancing, so also Nature ceases to function when she has made herself manifest to the Purusha or the Self. Nothing is more modest than Prakriti, when she becomes conscious that she has been seen by the Purusha. She does not again expose herself to the gaze of the Purusha.

Background to Samkhya Philosophy

In one of India's most renowned Scriptures-the *Shrimad Bhagavatam*-in chapter 25, verse 7 of the third canto, a saintly lady is speaking to her son. She addresses her son as "Lord", being fully aware that He is an Avatar-a descended manifestation of God.

She says: *nirvinna nitaram bhuman asat-indriya-tarshanat yena-sambhavya manena prapanna-andham tamah bhavo*

"I am so very much disgusted, oh my Lord, by the agitation of these temporary material senses; and on account of which I have fallen into the abyss of ignorance."

The lady's name was Devahuti, wife of the Yogi Kardama Muni, and her divine son was Lord Kapila.

Lord Kapila's reply to his mother constitutes what came to be known as *Samkhya Philosophy*. The word *Samkhya* in Sanskrit means *measure, counting, enumerating* or *categorising*. Lord Kapila enumerates the ways by which ordinary beings living in the world and bound by family ties, social commitments and economic entanglements can find the way back to their own true nature which is peace, love and unity with the divine.

Lord Kapila says in chapter 26, verse 1:-

"My dear mother, now I shall describe to you the different categories of the Absolute Truth, knowing which any person can be released from the influence of the Gunas."

Devahuti's condition is not unlike ours. It would seem that this condition of perplexity, confusion and frustration, brought about by the association with material elements and sensory objects, is independent of time, place and personality.

Devahuti, the mother of an Avatar, is seeking enlightenment, refuge, release from the clutches of the powerful Gunas, that seem to bind and entangle the being. Many of us feel as if we have fallen into a trap, a rut, an abyss we can't get out of.

".....I am disgusted, agitated, bewildered..." Devahuti says. Later on, in verse 29 she continues:-".....Please let me know what that system of yoga is and how many ways are there by which one may understand in truth that sublime yoga..."

The Bhagavatam is not the only place where *Samkhya* is mentioned. Thousands of years after the advent of Lord Kapila, we find Lord Krishna advising His disciple and friend Arjuna about Samkhya. In verse 39, chapter 2 of the Bhagavad Gita Krishna tells Arjuna:-

"I have declared to you the analytical knowledge of Samkhya Philosophy. Now listen to the knowledge of Karma Yoga...."

Much later, in chapter 18 verses 13 and 14 we get another glimpse of the depth of Samkhya Philosophy from Krishna's reference to it. He says:-

"Oh mighty armed Arjuna, learn from me the five factors which bring about the accomplishment of any action, as declared in Samkhya Philosophy....."

In chapter 5 verse 4 Krishna makes another reference to Samkhya Yoga as being equal to Karma and Bhakti Yoga combined.

He says:-

"Only the ignorant speak of Karma Yoga and Bhakti Yoga as being different from Samkhya Yoga...."

The narrator of the story of Devahuti and her divine son Kapila concludes by saying, in chapter 33, verse 1:-

"Thus Devahuti, the mother of Lord Kapila, and wife of Kardama Muni, became freed from all ignorance..... She offered her obeisances unto the Lord, the author of the principles of the Samkhya System of Philosophy, which is the background of liberation."

Some readers may be familiar with the concept of different "bodies" which are said to comprise the total human person. In terms of the physical, subtle, causal and spiritual bodies the Samkhya system can be seen to agree. If we group the first five elements into one, you have the physical or gross body. The Mind and the Intellect can be seen to comprise the subtle body. Similarly the ego is the causal body being the part that causes you to be and do and want the things you desire. And finally, the Self is your spiritual self, that part which is eternal and beyond the perception of the other three.

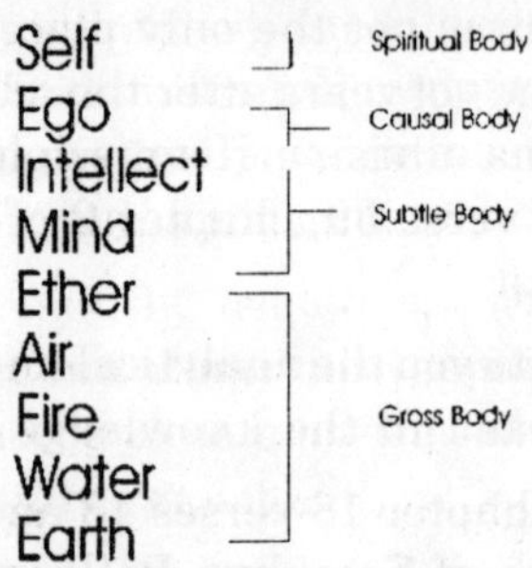

Universal Elemental Substances Grouped by Bodies

The Elements in Scalar Form

Presenting the elements in scalar form will make it possible to discuss various aspects of philosophical theory as well as map out a plan towards Self-knowledge.

The concept of "scale" is most useful in understanding how something works, what is its nature is and what can be predicted from it, its evolution and development. A scale has some sort of predictability. You can extrapolate information that may not be obvious from scattered data. A scale has a beginning and an end. You know where you are going if you follow a scale. And you can look back and see where you have been.

There are various kinds of scales. Some are linear; *i.e.* the progression from one step to another is constant. Some scales are exponential: the interval from one step to the next increases with each step. Some have a mixture of intervals, such as the musical scale, in which the intervals between steps are not even.

The subject of Samkhya Yoga Philosophy is the study of the scale of elemental substances which make up the human person. Indeed, Samkhya Philosophy in its fullness takes into account the details of the processes which precede the appearance of the human person in creation.

We are going to examine various aspects of the scale of elemental substances as presented by the Vedas and the Philosophy of Samkhya in particular. You will discover that we are dealing with a scale of universal importance. The "steps" of the scale will bring us to "ourselves" in the end. There are nine steps to this scale. The Self is the last stop.

The Elements in Brief

First of all let us just look at the elements themselves in brief in order to get acquainted with the elemental definitions according to the Vedic tradition.

Earth: All that is solid, has shape, form, solidity and mass. Characterised by the sense of smell. The grossest of human foods. The "body" of all planets and moons. All physical objects such as furniture, clothes, buildings, equipment, etc. Beings whose body is solely earth include rocks, gems, crystals, etc.

Water: Pure water has the quality of taste but no odour. Takes the shape of the container. Can move in and on the element earth. Holds the shape of earth objects. Is the cleanser and purifier. Another human food. Beings whose body is solely of water include rivers, lakes, snowflakes, oceans, etc. Water consumes (dissolves) earth.

Fire: Heat and light. Assists digestion. Gives objects their form and colour. Warms up the earth to assist in the growth of plants and vegetables. Its characteristic is sight. Beings whose body is essentially fire include suns and stars. Fire consumes both water and earth.

Air: Gives objects the quality of lightness. Its characteristic is touch, through which we sense temperature, shape and texture. It is the life-breath of all living things. Air supports fire, and penetrates both water and earth. Some beings whose body is solely air include Winds, Hurricanes, Tornadoes and other entities such as twirlies and twisties.

Ether: Known as the universal transformer. Above it is the subtle world, below it the gross world. Its characteristic is sound. It "contains" all the elements below it. We can appreciate it as "space" for our work. The Sanskrit for Ether is Akasha and is the realm which retains everything which has been spoken. When we have an "insight" we are accessing the ether- and it usually happens when we are very quiet and silent. The sense associated with space is listening. We would be wise to start "listening" to everything spoken as well as all other sounds that are made. Each sound contains valuable information about the situation in front of us. Ether is close to the next element above it (Mind). It is imperative that we "cross it" before we are able to tackle the mind itself. Beings whose body is essentially composed of ether are the so-called fairies, and other so-called "spiritual" essences or "guides".

Mind: The mind is characterised by thinking. Its objects of sense are thought, memories and imagination. The mind only contains mental image pictures of previously perceived sensations. The mind is larger than space in terms of dimension. The mind is usually divided into three areas in order to understand and discuss its function.

The top or "conscious" layer mind is called Buddhi and is that part of mind which can discriminate truth from untruth, it can cognize, constate and differentiate. Its operation is in the now and is the part which connects to the senses. There is cognition rather than recognition; seeing things as they are rather than comparing with what is already known, etc.

The middle or "subconscious" layer is called Manas and is responsible for the day-to-day housekeeping of our affairs. It compares, categorises, stores information for future use and can associate and compose new pictures made up of parts of previously perceived sensations.

The inner or "unconscious" layer of mind is called Chitta and is the seat of memory, un-finished business and the sense of purpose. This area of mind is not commonly accessible. It is, nonetheless, the basis for all our actions, decisions and inner so-called urges to do.

The senses associated with the Mind, then, are: Discrimination for the Buddhi; Association for the Manas; and Recall for the Chitta. Beings whose body consists of purely mind are called ghosts. It is quite possible that a mind which has been extremely attached to some physical object or person, lingers on or hovers around and can be "felt" by some people who have developed their subtle senses. The mind penetrates space, and all the other elements below it. It can "contain" all the other elements below it in "mental" form.

Intellect: This element is what has been commonly called intuition. It has nothing to do with so-called being smart or having a high IQ. It does not ordinarily function in the normal course of executing one's affairs because we ordinarily operate in the middle or subconscious layer of mind. To allow the Buddhi to operate, what is required is a connection with the senses and a falling still of the Manas. The intellect cannot be *forced* to function. It is always ready to function provided the obstructions are removed-*i.e.* the discursive manas and the agitation in the body. It is not easy to conceive of a being who is pure intelligence from our point of view because we are dealing with a substance which is subtler than thought-*i.e.* it cannot be thought about. We can only perceive the action of intelligence by what it does in a living organism including plants. Consider, for example how the body repairs itself when wounded or how species reproduce themselves. It might be presumed that if there are angels (such as guardian angels) these "beings" could be made from the substance of pure intelligence (our intuition).

Ego: This is the actual final frontier for us to cross. This is the operation of the Intelligence in response to memories held in the Chitta (memory). The substance of which ego is made is borrowed from the substance of the Self itself. We turn this substance into an image, as it were, and we call ourselves Somebody or Something. Once again, we cannot "see" the ego, because it is subtler than anything we have senses for. The essential substance of the ego is the clinging to an identity which is formed or acquired as a result of desire-the desire to be something or somebody. This substance is held in the sense of *being* this some one or some thing-it is not "imagined"-that would be a function of the mind. The mind cannot actually see the ego. The only way to "see" the ego is to "be" ourselves.

Self: Just as our body is a piece of the one earth, the one water, the one air, the one fire, the one space; so too our mind is of the one mind and our "self" is of the one Self. The closer we get to this realisation, the less we think of ourselves as separate beings, but part of the one being. WE ARE. Existence is our nature. The essence of our being is consciousness. Just pure consciousness. Fully aware, constant, unchanging, untouched by time and events (mind and world). Just witnessing time and space, and content in our own beingness, being totally happy, with nowhere to go and nothing to do. As soon as we desire to do something or go somewhere, then a vehicle (the ego and the rest) is immediately made available to us by our own will. We are the creators of our own life-if we want one.

What follows is an analysis of each element in association:

with a number of characteristics, function, properties, etc. Thirteen scales, labelled A to M have been defined in an attempt to present some of the rudimentary implications of Samkhya philosophy.

The Mind-Body Problem

I don't think that anyone actually denies the existence of these three latter elements-the Mind, the Intellect and the Ego. Opinions may differ, however, about their nature, constitution and definition. Consider for a moment, for example, the question: "Where is the mind located?" Modern psychologists and philosophers have discussed this question for a long time. Now consider the implication of the term "location". Location implies a position in space. But we see that mind is placed one rank higher than space in these ancient texts. And that implies, according to the hierarchical arrangement of the elements, that mind is finer than space.

In other words it is beyond the dimension of space and is therefore not contained in it or bound by it. And that is why we can't "find" it. Furthermore it is clear from the analysis of the senses we have dealt with so far, that none of the senses discussed could perceive mind since the perceptual density as defined earlier would be zero, which implies that it cannot be smelled, tasted, seen, felt or heard.

Mind and Intellect Differentiated

Also, according to the same analysis, the mind can be perceived by the intellect or at least by the sense organ which corresponds to the intellect. This organ is unknown to western science but it has a name in SANSKRIT called *buddhi* (Note: Buddha = one who is in constant touch with buddhi or "enlightened").

The working of our mind, our thoughts, ideas, mental speculations etc., can be "observed" by our intellect. In fact sometimes we check ourselves by saying "What I was thinking just a moment ago is wrong-I can see that now". Also when we are in a reverie, dreaming or "thinking", it is the intellect that pulls the mind back under control.

On the other hand we also experience the case of the mind not being able to "see" the intellect as when we are so much stuck in our own thoughts and opinions that we don't allow our intellect to function and hence we don't "see reason".

That the intellect is an element apart from the mind can best be appreciated if we think of the mind as the work-horse or the "computer" and the intellect as the driver or the computer programmer. The mind can't do anything by itself; it has to be directed. When it does apparently work by itself it is usually just engaged in fruitless circling thoughts that lead nowhere and accomplish nothing. But when the intellect controls the mind, that information is handled in a more organised fashion and the appropriate solution reached.

Material and Spiritual Qualities

Our quest is to realise ourselves as we are. What are we? We are constantly changing into something else are we not? Our bodies are changing, our habits are changing, our ideas and beliefs are changing, the way we feel about things and about ourselves is changing, and our roles are changing...

So is there anything about us that is not changing?

Spirit (or Truth) is defined as that which does not change in Time, Space or circumstance. We have come to appreciate this as the sense of being, attributable only to the Self-which is the only thing spiritual.

Let us consider what some of the characteristics of the spiritual realm are as compared to those in the material realm. We will find that we can function in both or in either depending on our consciousness.

The scriptures proclaim that the whole of the eight elements excluding the Self are "material"

We might make a distinction, however, between the upper three elements and the lower five and say that mind, intelligence and ego are made up of subtle substance (or matter) and the lower five are made of gross (physical) matter. But notwithstanding the subtlety or density of these "matters" they are all "material", the whole eight of them.

Qualities of the material world:

1. Pleasure decreases with repetition of stimulus.
2. Energy lost through activity, talking, singing, working, etc.
3. Love not possible without attraction, liking, pleasure, etc.
4. Sense of lack, incompleteness, separateness, longing for something higher, etc.
5. Everything decays and dies.
6. There is birth, death, disease, old age and suffering.

7. Lust for power, control, independence, knowledge and sensory gratification.
8. Take away from full and the remainder is not full.
9. Ruled by the Mind and the Ego and subject to Time and death.

Qualities of the Spiritual world:

1. Pleasure increases with repetition of stimulus.
2. The fabric of consciousness is Love, Unity, One-ness, Knowledge and Bliss.
3. Sense of completeness at each level although pleasure increases with each perception.
4. Sense of one-ness. Every one is different but the same at the same time.
5. No need for Mind-consciousness extends into "past" and "future", there is only "now".
6. Nothing dies. Everything is ever-fresh and new.
7. No decay, old age or death, nor birth or sickness.
8. Feeling independent and yet part of everything-being everything.
9. No energy loss through activity, sex, singing, dancing.
10. No hate, dislike, aversion, repulsion-these would be Self-contradictory (*i.e.* you would repel yourself).
11. Bliss is ever-increasing-although at any "time" you feel full.
12. Take away from full and the remainder is full.

Implications: In the material world (and be reminded that the "world" is defined by your own consciousness, not by externals) you look at something or someone beautiful and feel pleasure. The next time you look at the beloved object you feel less pleasure. The pleasure decreases each time and with the passage of time and circumstance. In contrast, in the spiritual world, each time you look at your beloved object you get an increase in pleasure and satisfaction. And each time you think that you could not possibly feel any more pleasure.

In the material world you get exhausted, even from lovemaking, eating, having a good time, etc. In the spiritual world no energy is lost (there is no friction and no resistance) and you can go on enjoying for ever without limit.

In the material world you have to undergo the process of birth, death, disease and old age, including all the suffering associated with each stage. In the spiritual world there is no birth and no death, no change and no decay. Everything is ever-fresh, new and variegated. You are not limited by time and space. You can be anything or anybody.

The sages have said that if we purify our mind of all desire, ambition, doubt, anxiety, fear, etc., we can perceive the spiritual world here and now. Remember that the higher you go up the scale of elements the greater is the field of operation of the substance comprising the element. So the Essential substance of the Self interpenetrates all the lower spheres. This means that Spirit is here where we are right now. It only takes a certain kind of looking at things. A certain kind of surrender and an acceptance of ourselves and everything that bless this creation with beauty, music, fragrance, etc.

We require to develop spiritual eyes in order to see the spiritual world. When our mind is purified in this way it can then reflect the spiritual world and becomes pure consciousness.

Transformation: The transition from Spiritual to Material takes place when we claim something for ourselves. For example, you fall in love with someone and everything is wonderful for some time. You are just lovers. Each is trying to please the other. You miss them terribly when they are not there to share with you. You want to share space, time, experiences, etc. Each time you see them the pleasure seems to increase. You are in bliss and ecstasy. Nothing else is quite as important as being with the beloved.

Then you begin to make plans for the future. There's the relations to consider, a house, money, the wedding, the church, the place to live, the school the children will go to, etc. You have now entered the material world where there are differences of opinion, tastes, personal history, personal likes and dislikes, ideas, beliefs, ingrained habitual mannerisms which are often hidden from your conscious awareness, etc. The love has gone.

Now you have a struggle to get your way and not appear an idiot in front of your friends and relatives. You have to show something of value in what you are doing. All the considerations about the other person have transferred to considerations about yourself and who you "are". The love you first experienced will never be the same. You have claimed it and have implicitly demanded that it remains yours in your own terms. You have made a claim you have no birthright to. It is a claim made by the acquired ego-identity. It is not Truth any more, but illusion. And there is no future in it.

The same thing happens when we make similar claims for "our" country, our property, our school, our team, our job, etc. The Scriptures say "claim nothing-everything belongs to the Lord". However the Scriptures also go on to say: "enjoy-heaven is spread all over the face of the earth". So the trick is to stay with the feeling of enjoyment and sharing, and not to make any move to claim for oneself that which belongs to the Self-the Self of everyone.

Self-Realisation: We have spoken about dimensions of experience which are made available through the various senses. We might ask now what are the dimensions of experience for a "Self"-realised person? The Vedas tell us that the nature of Self is "All knowing, all blissful and eternal". Thus the "self" who has "seen" God does not experience ignorance, does not experience death, and does not experience suffering. This has been manifested by many Saints, both in Western and Eastern religious history. When a Saint has achieved love of God (that's what makes him a "Saint") he no longer suffers, even if his body may be afflicted by physical pain or disease or tortured by other living entities. The Saint exhibits knowledge and wisdom which bypasses and surpasses the intellect. (There are many examples of Indian as well as western Sages and Saints who, although quite illiterate and uneducated, spoke very deep philosophical truths found only in the most advanced scriptures).

Other Philosophical Questions: Many philosophers and teachers have addressed the question of the nature of the "Self". The Samkhya philosophy makes it very clear that the Self may not be "studied" in the same sense that we may study the movements of planets, the process of growth, the chemical constitution of the "scientific" elements, etc.

Inspection of the scale of elements at once shows that the Self, being at the top of the scale, cannot be observed by elements below in the scale. That means that the mind cannot "think" of it; the Intellect cannot "reason" it; the ego cannot "identify" with it, and so on.

The Self is not Material: According to the Bhagavad Gita ("The Song of God") the Self is a Spiritual element and is eternally separate and different from "material" elements. We should appreciate by now that "element" implies actual substance, form, qualities, attributes, and so on.

Just as we appreciate that although we don't "see" air, nonetheless it has its own form, and it just happens that we can "sense" it by feeling etc. So too these other higher and subtler "substances" also have their form and can be "perceived" by the appropriate senses.

It would be naive, to say the least, then, to ask "Can you show me a self?" Even so, we can get a faint understanding of Self if we try to elevate our thinking as far as the element of Ego. At this level, you will recall the "sense" associated with Ego is the sense of being.

Now, if we fall very quiet within ourselves and try to contemplate on the sense of being, the question will gradually arise: "What am I, really?" And then we might notice that answers like: "I am an engineer, I am a father, a Christian, an Australian, a man, a woman" etc., etc., are not really what we are.

We should notice that these are simply labels, identifications, or acquired names that we came to adopt in the process of time. We might notice too, that we are, even without any of these designations. In other words being does not depend on the bodily or mental designation. The scriptures refer to these designations as "false ego".

By implication, then, we can see that Self transcends the sense of "I am something". Also, according to the logic we have developed so far, perception of any element takes place from a level higher than that which is perceived. So when a person "sees" that what he has been labelling himself with is false ego, he is said to be self-realised.

Divine Love and Divine Perception: If we pursue the logic we have developed in terms of the scale of perceptivity, we see that we would require another element which can "see" all the so-called "selves". That element has been called by various names in the Scriptures; but whether we say Absolute, or God, or Supreme Being, Paramatma, etc., the essence of the element in question is the same: It is that substance of which all the separated "selves" are its component parts.

Just as we might think of our body and soul as the difference between matter and spirit in us, so too all the souls together are the "body" of the Supreme Soul or Absolute, God, Paramatma, Supreme Being. And as our souls are spirit, so it is that God is only spirit. His entire body is constituted of all the Selves that we are.

The Scriptures say: that "perception" which takes place from the level of the Absolute or God is called Divine Love. So, when God "looks" at us (the "selves"), we experience His love.

The experience of Divine Love is not sensory in the usual sense. Because it is not material. Individuals who experience love of God won't be able to express it in words and often don't try. Symptomatic of someone who is experiencing love of God is the fact that he feels so completely satisfied that he has no desire for any of the lower, material pleasures. He may still eat, sleep, drink, talk, and live in the world (of sense) but he has no hankering for anything (no Ahankara), he does not lament for the loss of worldly goods and relationships etc. He is what's called "Self-satisfied".

In other words his "self" is satisfied. This shows why an individual who does not have love of God is not completely satisfied. His satisfaction may only pertain to the body, the mind, the intellect and the ego; but unless he is HIMSELF satisfied (unless God perceives him) he will not feel fully satisfied.

Example—On the Question of Pollution and Disease: Just a quick look at the scale of elements should show that pollution and disease can occur at any level. Considering the first five as a group, we may note that a room or a house in disarray, can be explained in terms of the scatter of physical objects (earth) in the wrong places (space). The room or house may not have the necessary amount of light, ventilation, heating, etc. We might find that earth (dust) lies where there should be light (fire). That is to say, the furniture may need a polish and a shine.

What of water and air? If these exceed their proper balance you may end up with dampness or drafts.

Of course we can pollute the *outside* of buildings with rubbish, sewerage, fumes, smoke, and clutter. Each individual contributes to space and air pollution; and it all has to do with the amount of attention one has when performing an action. Everything we do can and should be done with care for the environment. Because the environment is not ours to clutter and pollute. When we do we poison ourselves and reduce our chances of becoming free and conscious.

We can also pollute the inside of our own body-again, by the inappropriate amounts of each of the elements. Over-eating or drinking are of course the most likely things we do. We often consume too much of the wrong kind of food just for the sake of convenience and time-saving. But are we saving time really? And what do we do with the time saved? We could live a lot longer if we took the time to prepare our meals properly and eat slowly and with thanksgiving in mind, rather than with the thought that we have to get it over with because we have more important things to do.

To be conscious of the appropriate amount of anything we consume, only requires to be in touch with the elements at the time. Our sense of touch gets us in touch with the Air; our sense of smell with the Earth. Our sense of taste with the Water, the sense of sight with Light, and the sense of hearing with Space. It is just a question of being in touch at the time. Our body will tell us immediately if we are about to do something inappropriate.

Using the senses properly will bring about enjoyment, whereas ignorance will bring about pain and suffering-although at the time it may appear to be enjoyment. Ignorance is never bliss.

The diseases resulting from an abuse of the body are well known: lethargy, congestion, indigestion, breathing problems, headaches, tendency to sleep, inability to think, lack of energy, etc.

If we take the level of Mind to assess the kinds of pollution and disease which can occur, we will find that we actually do a lot of damage, even when we do not wish or intend to. Remember that the mind interfaces with the world via the senses. The senses are thus responsible for the contents of the mind.

How we use the senses determines whether they are our friends or our enemies. If I look at some object of beauty, I am nourished and feel uplifted inside. The same eyes can look at something unpleasant or even specially designed to arouse my lower nature in a way that may encourage and develop greed, lust and anger.

It is not to say that we must close our eyes to what is happening; that would be ignorance. But we do need to develop the wisdom (the function of Buddhi) to know the process of sensing accompanied by thinking. I should be able to look at the unpleasant object and arrest any thought about it, *i.e.*, I should become immediately aware of the process of thought arising in me and proceed with caution, knowing that the thought may one day give vent to an action which I am sure to regret. Thus I should practice becoming aware of the process of thinking arising from looking.

Similar remarks could be made about the other senses. The sensing is one thing. The thinking which arises out of the sensing is quite another. For example I have just tasted a piece of delicious cake that someone has offered.

Now, I have the choice to go with the thought "I think I will have another slice" or "That was delicious thank you", and finish there. How many times do we catch ourselves thinking

"I might as well go along with it", just to be socially accepted, or to try and repeat the pleasure!

Example—Some Hints on Daily Practices

On Retiring: Be sure to bring the three bodies into alignment before retiring (The Gross body = physical body; The Subtle Body = the mind; The Causal Body = the emotions). The Manas will most likely keep on moving unless subdued by the intellect (Buddhi) to reside with the physical body.

Routine (Brief): Sit or lie down and be sure to actually come in touch with the entire surface of your physical body, through the sense of touch. Follow this with coming in touch with the space immediately surrounding you through hearing. Sustain these two whilst running your attention up and down the spinal column a few times. Finally, maintain the attention on your physical body and lie down to rest.

Routine (Extended): Be sure to "wind down" on your last activity and properly and completely bring it to a close. Do a few cycles of coming in touch. Go over the main activities of the day and bring them to rest in the mind. Put the whole day behind you and settle down with a passage from Scripture. (Alternately, listen to a short passage of conscious music). Then carry out the brief routine above.

On Decision Making: Every day you will be faced with decisions to make. Often these take the form of choosing between two or more possible lines of action. The criterion should be *what is needed by the situation* at the time (the object of sense of the intellect). Watch for that initial reaction in the mind which enters as a result of habit and do not immediately respond to it. Upon cool examination of the situation it may well be that it fits.

If there is any tinge of personal opinion or personal gain in the initial reaction, drop it immediately. Decide for the best for everyone. Be detached but caring; concerned but not overwhelming. Consider the long term outcome as well as the immediate conditions. Aim for a movement upward on the table (the greater good), for everybody concerned, including yourself.

On Choosing: More often than not the choice you will be presented with is likely to be a choice between alternatives. Remember that real choice consists in deliberately and consciously doing or not doing something.

Choice is singular, about one thing; not two or more. Choosing between two or more things can be habitual; choosing what is more familiar, more pleasant, and more personally beneficial. This is the work of Manas and habit. It does not necessarily constitute evolution and re-finement. To choose *not to do* something which habit tells you strongly to do is conscious work on self and results in purification of the Ego.

Classical Samkhya

The philosophical speculations found in what I have called the *Atreya-Tantra* are a pre-classical Samkhya text, a precursor of the more famous *Samkhya Karika of Ishvara Krishna*.

The latter is an exposition of the system as taught by the philosopher Kapila (18) and is widely considered to be is the most authoritative record of Samkhya views. Kapila is supposed to have lived sometime before 200 AD at the end of the 'Epic' period of Indian cultural history. D Chattopadhyaya advances the theory that the name Kapila is a derivative of Kapila (long 'a' as in father) (19) which indicates that at least one of the original Samkhya adepts was female.

This would be an important characteristic of the view. It may indeed be possible to identify, in the Samkhya views, elements that represent a female sensibility. For example the role of (female) nature and its abhorence of warfare. There is a passage in the *Mahabharata* concerning Pancashikha, one of many great Samkhya masters. It relates that:

Pancashikha became a disciple of Asuri. He lived on human
milk. There was a certain brahmani called Kapila. She was the
wife of Asuri. Pancashikha was accepted by her as a son and he

used to suck her breasts. In consequence of this, he came to be

known as the son of Kapila and his understanding became fixed on

Brahman. All this, about the circumstances of his birth and

those that lead to his becoming the son of Kapila, was told to

me by a divine sage.

Pancashikha, whose name means, "five-crested one", is himself a bit of a strange character and may, like his teacher Asuri, have been of the race of elder gods or, as the Brahmins called them demons, who ruled before the coming of the Aryans. (see *Demon Doctrine of Tantra*).

The reference to his diet of human milk is also very strange and suggests demonic origins, unless we are to suppose he became a student whilst still a baby, not such a far fetched idea in the Indian tradition.. Suckling is a common metaphor for the flow of life giving knowledge, although the usually an ordinary cow and its calf is the subject. Indeed, the fate of the cow is very dear to Kapila's heart as we shall see in one of the stories related below. This suckling metaphor also suggests to me some of the later tantrik practices in which spiritual knowledge is passed to the initiate via bodily fluids.

One of the strongest pieces of evidence for the dating of the Samkhya view is anecdote found in the Buddhist tradition concerning the birthplace of the Buddha, which was named after Kapila. The Buddha died in 544BCE, so the system of Kapila must have been formulated at least before that time in about the seventh century before the present era.

Thus the town named after Kapila rejoiced with its surrounding territory at the prosperous birth of the prince.

These lines are from the *Buddha-Carita,* a 'biography' of the Buddha, written during the reign of the potentate Kanishka and therefore lst-2nd century. This same document tells of the apprenticeship of the Boddhisattva, and how he visited the hermitage of Arada, another adept of the Samkhya lineage. The Buddha-to-be is supposed to have repudiated Arada with these words:

> *I have listened to this doctrine of yours, which grows more subtle and auspicious in its successive stages, but I consider it not to lead to final beatitude, since the 'knower of the field' is not abandoned. Thus he was not satisfied on learning the doctrine of Arada and, discerning that it was incomplete, he turned away from there.*

The above passages along with others indicate that the Buddha studied Samkhya philosophy and that his own view is in fact a development from a Samkhya base. And indeed the Samkhya found in the *Atreya Tantra,* contains notable passages of what could be anti-Buddhist critique.

Samkhya is clearly the name of a very ancient Indian philosophical view; one of two that superceded the Upanishads. According to at least one modern-day commentator it was a reaction against the growing idealism of the Upanishadic thinkers.

Samkhya is commonly designated as a *Dualist* philosophy. Dualism is normally thought to postulate an universe comprised of two, largely independent principles such as Self (purusha) and Matter (prakriti). The kind of Dualism found in these texts is quite different to the variety more familiar in western philosophy from the works of Rene Descartes.

The physician Caraka uses the term Samkhya several times; either as an epithet of a knowledgeable teacher and along with Yoga as the name for a system of philosophy.

The word Samkhya derives from the Sanskrit root 'Khya' meaning to discriminate or number. We can infer from this that there was obviously a lot more to Samkhya than Dualism. The scholar Edgerton suggested that Samkhya was the Reasoning or *Gnostic* school.

The first verse of Ishvara Krishna's *Samkhya-Karika* discloses the principal or sole purpose of the system. This is stated to be the relief of humanity from suffering and pain:

Because of torment by three kinds of pain (there arises) the desire to know the means of removing it. If because of the obvious (means of removing it) this (desire) seems superfluous; it is not so, for these are neither absolutely complete nor abiding.

This concern with the relief of humanity from pain was an interest shared by Ayurvedic physicians, some of whom like the Siddhas, were also Tantrik adepts. This again is a point of contrast between eastern and western intellectual tradition. For example, the Dualist philosopher Descartes, begins his discourse on method as a disinterested inquiry.

Then with the advent of embodied diseases this became an obstacle (vighnabhuta) to wealth, sacred vows, the study of the Veda, meditation and fasting. Freedom from disease is the chief root of Duty (Dharma), Prosperity (Artha), Pleasure (Kama) and Final Liberation (Moksha). Diseases will take away the best part of this and life (itself).

If you take away the first of these (duty), you have the three purposes of life as expressed in the tantrik tradition.

The Samkhya philosophers dealt with the notion of pain under three categories.

i. The intrinsic (adhyatmika)

ii. The extrinsic (adhibhautika)

iii. The divine or superhuman (adhidaivika).

Of these, the intrinsic is two-fold, bodily and mental. Bodily pain is caused by the disorder of the several humours, wind, bile and phlegm.

Mental pain is due to desire, wrath, avarice, affection, fear, envy, grief, and the non-perception of particular objects. These later are in fact 'kleshas', a medical term meaning pain. In the Nath tradition, the klesha are given as five in number:

The five pain bearing obstructions the root causes of trouble and strife ignorance, ego, revulsion, attachment and clinging to life. (Tantra Magick : 54)

The AMOOKOS adepts trace these back to the *Linga Purana:*

> *By means of the fivefold bonds called klesha, Shankara (Shiva) binds the pashus. On being served well by devotion, he alone is their redemmer. The five kleshas that have become fetters are ignorance, ego, revulsion, clinging to life. (Linga Purana II.9)*

The kleshas and other pains above are all of the intrinsic variety on account of their being amenable to internal remedies.

Pains amenable to external remedies are two-fold: extrinsic and superhuman. The extrinsic are caused by men beasts, birds, reptiles and inanimate things; and the superhuman ones are due to the evil influence of planets and the various elementals.

An objection is raised: '.... Hundreds of remedies for bodily pain are laid down by eminent physicians; for mental pains also we have easy remedies in the shape of the attainment of the objects of enjoyment such as women, desirable food and drink, unguents, dress, ornaments, and the like.

Similarly, of extrinsic pains we have easy remedies such as expert knowledge of moral and political science, residence in safe places etc. In the same manner of superhuman troubles we have remedies in the shape of charms, incantations, and the rest. (The Reply).... though easily available, the obvious means do not effect absolute and final removal of pain. Consequently the present enquiry [for spiritual remedy] is not superfluous.

The tantrik or spiritual physician wants to cure the pain of human existence permanently. Medicine cannot solve these kinds of issues but is can buy you a little time with which to think about them. There is, after all, nothing quite like illness to fix you in the mundane world of matter. People say that illnesses such as headache or fever tend to return.

The causes of disease in this instance an imbalance of the humours (dosa), return again. Therefore the appropriate remedy for a disease, whether physical or spiritual, will be one that looks at what happened in the past. 'As a dam is made in order to hold back the swelling river waters, which had previously damaged the cornfield.' In the same way, he who intends to administer a therapy to destroy future pain depends upon having seen the early development and history of the disease. The cycle of disease is then extinguished and pleasure arises from the therapy.

These doctrines treats physical and mental trauma on an equal footing, implying that one cannot progress to the contemplation of spiritual things without also thinking about your physical fitness. To take a very simple example; a person could not maintain one of the postures (asanas) that are an integral part of meditation without a moderate degree of good health.

The relationship between health and sanctity is a dialectical one-one is nourished by the other. Samkhya philosophy is sometimes said to be a radical departure from the formalism of the ancient Vedic texts. The second aphorism of the Samkhya Karika rejects the Vedic sacrificial religion in favour of its particular path of knowledge.

The revealed (means) are like (medicine) for they are connected with impurity, destruction and excess. A contrary method is better and this consists in a discriminative knowledge of the manifest (forms of matter), The Unmanifest (Prakrti or primal matter) and the knowing-self (jna).

We can agree, to some extent at least, with this condemnation. Even medicine does involve 'impurity, destruction and excess`, whether this be in the form of the animal sacrifice of Vedic religion or even the use of animal products in the medical pharmacopeia.. In our own time, there has been renewed criticism of Chinese traditional medicine which makes use of exotics like bear's bile. And our own western system of medicine, called allopathic in India, makes a great, and to some unacceptable use, of vivisection.

In ancient India, the establishment of veterinary dispensaries at about this time, indicated a growing concern for the welfare of animals. It is surely significant that in the popular mind, the celebrated veterinarian Shalihotra is supposed to have been Kapila's son. There is an extremely interesting dialogue in the Mahabharata, which throws some light on Kapila's attitude to Vedic sacrifice. When the deity Tashtri came to the palace of king Nahusha, the latter for hospitality was on the point of killing a cow.

Beholding that cow tied for slaughter Kapila of liberal soul, cried 'Alas ye Vedas!' At that time a Rishi called Syumarashmi entered (by yogic power) the form of that cow and addressed Kapila thus: 'If the Vedas be authoritative, whence have those other duties concerning harmlessness to all creatures come to be regarded as important?'

It is obvious that this dialogue occurs at a time when the notion of non harming (ahimsa) has already many adherents in ancient India, not only amongst early Buddhist but also within Hinduism.

The protagonist here is pointing out the apparent contradiction between Hinduism with its promulgation of non-harm to all creatures and the older and still venerated, indeed supposedly authoitattive *Vedas*, that call for many bloody sacrifices.

Kapila replies: 'I do not censure the Vedas, I do not wish to say anything in derogation of them. However I have heard that the different courses of duty that are laid down for the different modes of life, all lead to the same end. So both to perform and not perform acts are both Vedic declarations. When the scripture appears thus, the strength or weakness of particular declarations must be very difficult to ascertain. If thou knowest of any course of duty which is superior, then tell me about it.'

Syumarashmi replies by reaffirming the ancient tradition of sacrifice, in which the inferior animals, human beings, trees and herbs all wish for heaven. There is no means, other than sacrifice, by which they can fulfil that desire.

Kapila rejoins: 'The fruits attainable by acts are finite instead of eternal. Ascetics, by adopting self-restraint and
tranquillity attain to God (Brahman) through the path of

knowledge. They are freed from the influence of all pairs of

opposites. Wisdom is theirs. Cleansed they are from every sin.

Transcending grief and freed from impurity (rajas), theirs are

acquisitions that are eternal. When the high end that is these

men's, is within reach, what need has one for practicing the

duties of the domestic mode of life?'

Syumarashmi replies that if indeed, that be the highest object of acquisition if that be truly the highest end, then the importance of the domestic mode of life becomes obvious, because without this, no other mode of life would be possible! Indeed as all living creatures are able to live in consequence of their dependence on their respective mothers, in the same way the three other modes of life exist in consequence of their dependence upon the domestic lifestyle; that householder who leads the life of domesticity, performs sacrifices and practices penances.

Your doctrine is false, though dressed in the colours of truth. It subverts the real purport of the declarations of the Vedas and must have been introduced by learned men reft of prosperity and eaten up by idleness.

The Brahmin who performs sacrifice according to the declarations of the Vedas is never seduced by sin. Through sacrifice, such a person attains the highest regions of felicity, along with the animals he has slain in those sacrifices. And by gratifying his own wishes, succeeds in gratifying the wishes of those animals.'

Kapila replies: that if acts are obligatory then the Darsha (fortnightly oblation), Paurnamasa (full-moon oblation), Agnihotra (fire-sacrifice) and Caturmasa (Fourth month sacrifice) should be sufficient; why is there any need to partake in acts involving cruelty? One who has no fear for any creature and from whom no creature hath any fear and who constitutes himself the soul of all creatures, should be known as a Brahmin. Without having acquired purity of heart the result of pious acts, one remains a foolish man and will not be a genuine Brahmin no matter who ones teachers are.

The ignorant man desire fruits of a very different kind. He is unable to practice even a small part of the good conduct recommended since remote times. Such foolish men don't really understand or appreciate the Vedic sanctions. Besides which as regards sacrifices, it is extremely difficult to get all the details correct and so the results can only be of limited value.'

At this point Syumarashmi capitulates and reveals his true identity. Acknowledging Kapila as his teacher she asks: 'The Vedas countenance and discountenance acts. Whence then is their authority when their declarations contradict each other?'

Kapila replies that the path of Yoga is the best path because its results are visible in this world. It is only through ignorance that a contradiction is seen between this view and that of the Vedas.

Kapila claims to be recapturing the visions of the ancient Vedic seers. 'There were,' he says, ' in days of yore, many men leading lives of domesticity. They were thoroughly devoted to their own duties. There were also many kings of the same qualification as well as Brahmins, all devoted to yoga.

They behaved equally towards all creatures and were endowed with perfect sincerity, contentment and certainty of knowledge. Manifest were the rewards of their righteousness and pure were they in behaviour and heart'. 'Acts ' he says 'only cleanse the body. The highest end is knowledge itself. When one possesses this, all the faults of the heart are cured. When one is at peace with God (Brahman), one becomes established in knowledge, one obtains benevolence, forgiveness, tranquillity, compassion truthfulness, candour and devotion.'

It is clear that in the popular mind and in the mind of the Samkhya philosophers, there was never any thought of heresy. Their aim was always to recapture what was to them the highest aspect of Vedic religion which had been lost. They placed their complete faith in the path of knowledge which to them was worth a great deal more then animal sacrifices and Soma (vajapeya).

A thousand horse sacrifices, one hundred Vajapeya are not worth one sixteenth of the merit of yoga (52)

It is out of this philosophy that many Tantrik ideas developed. It is not uncommon for tantrism to be viewed as anti establishment. But early pieces of discussion such as the above show that tantriks need not have shared this perception of themselves. They may indeed have thought themselves truer to the spirit of the tradition out of which they grew.

2

The Samkhya Aphorisms of Kapila

BOOK I

a. Salutation to the illustrious sage, Kapila!

b. Well, the great sage, Kapila, desirous of raising the world [from the Slough of Despond in which he found it sunk], perceiving that the knowledge of the *excellence* of any fruit, through the desire [which this excites] for the fruit, is a cause of people's betaking themselves to the means [adapted to the attainment of the fruit], declares [as follows] the excellence of the fruit [which he would urge our striving to obtain]:

Aph. 1.* Well, the complete cessation of pain [which is] of three kinds is the complete end of man.

a. The word 'well' serves as a benediction; [the particle *atha* being regarded as an auspicious one].

b. By saying that the complete cessation of pain, which is of three kinds,—*viz.*, (1) due to one's self (*adhyatmika*), (2) due to products of the elements (*adhibhautika*), and (3) due to supernatural causes (*adhidaivika*),—is the *complete* end of man, he means to say that it is the *chief* end of man, among the four human aims, [*viz.*, merit, wealth, pleasure, and *liberation* (see *Sahitya-darpana*, § 2)]; because the three are transitory, whereas liberation is *not* transitory: such is the state of the case.

c. But then, let it be that the above-mentionend cessation [of all the three kinds of pain] is the complete end of man; still, what reason is there for betaking one's self to a doctrinal system which is the cause of a knowledge of the truth, in the shape of the knowledge of the difference between Nature and Soul, when there are *easy* remedies for bodily pains, *viz.*, drugs, &c., and remedies for mental pains, *viz.*, beautiful women and delicate food, &c., and remedies for pains due to products of the elements, *viz.*, the residing in impregnable localities, &c., as is enjoined in the institutes of polity, and remedies for pains due to supernatural causes, *viz.*, gems [such as possess marvellous prophylactic properties], and spells, and herbs of mighty power, &c.; and when [on the other hand], since it is hard to get one to grapple with that very difficult knowledge of truth which can be perfected only by the toil of many successive births, it must be still *more* hard to get one to betake himself to the doctrinal system [which treats of the knowledge in question]? Therefore [*i.e.*, seeing that this may be asked] he declares [as follows]:

Aph. 2.* The effectuation of this [complete cessation of pain] is not [to be expected] by means of the visible [such as wealth, &c.]; for we see [on the loss of wealth, &c.,] the restoration [of the misery and evil,] after [its temporary] cessation.

a. 'The visible,' in the shape of the drugs, &c., above-mentioned [§ 1. *c.*].

b. 'The effectuation of this,' *i.e.*, the effectuation of the complete cessation of pain.

c. *Why* is it not [to be thus effected]? Because, after the cessation (the cessation of pain is understood), we see its restoration, the springing up again of pain in general, [from whichever of its three sources (§ 1. *b.*)].

d. The state of the matter is this: not by the expedients above-mentioned is there such a removal of pain, that no pain arises thereafter; for, when, by this or that expedient, this or that pain has been destroyed, we see other pains springing up. Therefore, though it be *not* easy [§ 1. *c.*], the knowledge of truth [as a complete remedy] *is* to be desired.

e. But then, grant that *future* pain is not debarred by drugs, &c., [employed to remove present pain], still, by again and again obviating it [as often as it presents itself], there may be the cessation of *future* pain, also. This doubt he states [as follows]:

Aph. 3.* [Let us consider the doubt] that the soul's desire [the cessation of pain, may result] from exertions for the obviation [of pain], as is the case with the obviation of daily hunger

a. When pain shall arise [let us suppose one to argue] then it is to be obviated; and thus there is the soul's desire, the cessation of pain; just as one should eat, when there is hunger; and thus there is the soul's desire of the eater, *viz.*, the cessation of hunger. In regard to this [doubt] he states the recognized decision:

Aph. 4.* This [method of palliatives (§ 3)] is to be rejected by those who are versed in evidence; because it is not everywhere possible [to employ it at all], and because, even if this *were* possible, there would be an impossibility as regards [ensuring] the perfect fitness [of the agents employed].

a. For there are not physicians, &c., in every place and at all times; and [to rely on physicians, &c., would not be advisable], even if there were the possibility,—*i.e.*, even if these *were* [always at hand], since physicians are not *perfect* [in their art];—for pain cannot with certainty be got rid of by means of physicians, &c., with their drugs, &c. Moreover, when corporeal pain has departed, there may still be that which is *mental*, &c.; so that there is not [under such circumstances], in every respect, liberation from pain. For these reasons, *such* a soul's aim [as that which contents itself with temporary palliatives] is to be rejected by those who are versed in evidence, [*i.e.*, who are acquainted with authoritative treatises].

b. He mentions another proof [of his assertion]:

Aph. 5.* Also [an inferior method ought not to be adopted] because of the preeminence of Liberation [as proved] by the text [of Scripture declaratory] of its preeminence above all else.

a. One ought not to endeavour after the removal of this or that pain by these and those expedients [§ 1. *c.*]; since Liberation (*moksha*), by being eternal, is transcendent as a remover of all pains. Moreover, one ought to endeavour only after the knowledge of truth, which is the means thereof [*i.e.*, of Liberation]; because the Scripture tells its preeminence above all [other objects of endeavour], in the text: 'There is nothing beyond the gaining of Soul, [with the utter exclusion of pain].'

b. But then [it may be suggested], when you say *liberation*, we understand you to mean from *bondage*. And is that bondage essential? Or is it adventitious? In the former case, it is incapable of destruction; if it come under the latter head, it will perish of itself, [like any other adventitious and, therefore, transitory thing]. What have we to do with your 'knowledge of truth,' then? To this he replies [as follows]:

Aph. 6.* And there is no difference between the two.

a. There is no difference in the applicability of liberation, on either of the suppositions, that the bondage is essential, and that it is adventitious, [supposing it were either (see § 19. *b*.)]. That is to say, we can tell both how the bondage takes place, and how the liberation takes place.

b. Now, with the view of demonstrating [the real nature of] Bondage and Liberation, he declares, exclusively, in the first place, the objections to Bondage's being *essential* [§ 5. *b*.]:

Aph. 7.* There would be no rule in the enjoining of means for the liberation of one bound *essentially*.

a. Since Liberation has been stated [§ 1] to result from the complete cessation of pain, [it follows that] Bondage is the junction of pain; and this is not *essential* in man. For, if that were the case, then there would be no rule, *i.e.*, no fitness, in the Scriptural or legal injunction of means for liberation: such is what must be supplied, [to complete the aphorism]. Because, to explain our meaning [by an illustration], *fire* cannot be liberated from its *heat*, which is eseential to it; since that which is *essential* exists as long as the substance exists.

b. And it has been declared in the Divine Song [the *Ishwara-gita*,]: 'If the soul were essentially foul, or impure, or changeable, then its liberation could not take place even through hundreds of successive births.'

c. [Since some one may be disposed to say] '*Grant* that there is no fitness [in the Scriptural and legal injunctions, (§ 7. *a*.)], what have we to do with *that*?' Therefore he declares [as follows]:

Aph. 8.* Since an essential nature is imperishable, unauthoritativness, betokened by impracticableness, [would be chargeable against the Scripture, if pain were essential to humanity].

a. That is to say: since the essential nature of anything is imperishable, *i.e.*, endures as long as the thing itself, it would follow [on the supposition that pain is essential to humanity], that, since Liberation is *impossible*, the Scripture which enjoins the means for its attainment is a false authority, inasmuch as it is impracticable [in its injunctions. And this is out of the question; Scripture being assumed, here, as in all the others of the six systems, to be an exact measure of truth].

b. But then [some one may say], let it *be* an injunction [to use means for the attainment of an unattainable object], on the mere strength of Scripture; [and, since Scripture is an unquestionable authority, we may be excused from asking or answering the question, *why* the injunction is given]. To this he replies [as follows]:

Aph. 9.* There is no rule, where something impossible is enjoined: though it *be* enjoined, it is no injunction.

a. There can be no fitness, or propriety, in an injunction with a view to an impossible fruit; seeing that, though something be enjoined, or ordered [to be effected] by means that are impracticable, this is no injunction at all, but only the *semblance* of an injunction; because it stands to reason, that not even the *Veda* can make one see sense in an absurdity: such is the meaning.

b. Here he comes upon a doubt:

Aph. 10.* If [some one says] as in the case of white cloth, or of a seed, [something essential may be not irremovable, then he will find his answer in the next aphorism].

a. But then [the doubter is supposed to argue], the destruction even of what is essential [in spite of what is stated under *s* seen; as, for example, the essential whiteness of white cloth is removed by dyeing, and the essential power of germination in a seed is removed by fire. Therefore, according to the analogy of the white cloth and the seed, it is possible that there should be the removal of the bondage of the soul, even though it *were* essential. So, too, there may be [without any impropriety] the enjoinment of the means thereof. Well, *if* [any one argues thus], such is the meaning [of the aphorism, to which he proceeds to reply].

b. He declares [the real state of the case, with reference to the doubt just raised]:

Aph. 11.* Since both perceptibleness and [subsequent] non-perceptibleness may belong to some power [which is indestructible], it is not something *impracticable* that is enjoined, [when one is directed to render some indestructible* power imperceptible].

a. In regard even to the two examples above-mentioned [§ 10], people do not give an injunction for [the positive destruction of]** something essential, which is indestructible [§ 8]. Why [do we say this]? Because, in these two instances of the perceptibleness and non-perceptibleness of a power [the powers, namely, of appearing white and of germinating (see § 10. *a*.)], there are merely the manifestation and [afterwards] the *hiding* of the whiteness, &c., but not the *removal* of the whiteness, or of the power of germination; because, that is to say, the whiteness of the dyed cloth and the germinating power of the roasted seed can again be brought out by the processes of the bleacher, &c., [in the case of the dyed cloth], and by the will of the *Yogi*, [the possessor of supernatural powers, in the case of the roasted seed], &c.

b. Having thus disproved the notion that bondage is *essential* [to man], wishing to disprove also the notion that it is the result of some [adherent] *cause*, be rejects the [various supposable] causes, *viz*., Time, &c.:

Aph. 12.* Not from connexion with *time* [does bondage befall the soul]; because this, all-pervading and eternal, is [eternally] associated with *all*, [and not with those alone who are in bondage].

a. The bondage of man is not caused by *time*; because [if that were the case,] there could be no such separation as that of the *liberated* and *unliberated*; because time, which applies to everything, and is eternal, is at all times associated with all men, [and must, therefore, bring *all* into bondage, if any].

Aph. 13.* Nor [does bondage arise] from connexion with *place*, either, for the same [reason]

a. That is to say: bondage does not arise from connexion with *place*. Why? 'For the same reason,' *i.e.*, for that stated in the preceding aphorism, *viz*., that, since it [*viz*., place] is connected with *all* men, whether liberated or not liberated, bondage would [in *that* case] befall the *liberated*, also.

Aph. 14.* Nor [does the bondage of the soul arise] from its being conditioned [by its standing among circumstances that clog it by limiting it]; because *that* is the fact in regard to [not the soul, but] the *body*.

a. By 'condition' we mean the being in the shape of a sort of association. The bondage [of the soul] does not arise from *that*; because *that* is the property of the *body* [and not of the soul]; because, that is to say, bondage might befall even the liberated [which is impossible], if that which is the fact in regard to another could occasion the bondage of one quite different.

b. But then [some one might say], *let* this conditioned state belong to the soul. On this point [to prevent mistakes], he declares:

Aph. 15.* Because this soul is [unassociated with any conditions or circumstances that could serve as its bonds, it is] absolute.

a. The word *iti* here shows that it [*i.e.*, the assertion conveyed in the aphorism] is a *reason*; the construction with the preceding aphorism being this, that, *since* the soul is unassociated, it belongs only to the body to be conditioned.

Aph. 16.* Nor [does the bondage of soul arise] from any work; because [works are] the property of another [*viz*., the mind], and because it [the bondage] would be eternal, [if the case were as you imagine].

a. That is to say: moreover, the bondage of the soul does not arise from any work, whether enjoined or forbidden; because works are the property of another, *i.e.*, not the property of the soul [but of the mind]. And, if, through a property of another, the bondage of one quite distinct could take place, then bondage might befall even the liberated, [through some acts of some one else].

b. But then [some one may say], this objection does not apply, if we hold that bondage may arise from the acts of the *associate* [*viz.*, the mental organ]: so, with allusion to this, he states another reason; 'and because it would be eternal,' *i.e.*, because bondage, in the shape of connexion with pain, would occur [where it does not,] even in such cases as the universal dissolutions [of the phenomenal universe, including the mental organ, but *not* the soul].

c. But then [some one may say], if that be the case, then let the bondage, too, in the shape of connexion with pain, belong [not to the *soul*, but] to the *mind* alone, in accordance with the principle that it have the same locus as the works [to which it is due]; and, since it is an established point that pain is an affection of the *mind*, why is bondage [*i.e.*, connexion with pain] assumed of the *soul*, also? With reference to this doubt, he declares [as follows]:

Aph. 17.* If it were the property of any other, then there could not be diverse experience.

a. If bondage, in the shape of connexion with pain, were the property of another, *i.e.*, a property of the *mind*, there could be no such thing as diverse experience; there could be no such different experience as one man's experiencing pain, and another man's not: [for, it must be remembered, it is not in point of *mind*, but of *soul*, that men are held, by Kapila, to be numerically different]. Therefore, it must be admitted that pain is connected with the soul, also. And this [pain that belongs to the soul] is in the shape merely of a *reflexion* of the pain [that attaches to its attendant organism]; and this reflexion is of its *own* attendant [organism] only; so that there is no undue result [deducible from our theory].

b. He rejects also the notion that Nature *(prakriti)* is *directly* the cause of bondage:

Aph. 18.* If [you say that the soul's bondage arises] from Nature, as its cause, [then I say] 'no;' [because] that, also, is a dependent thing.

a. But then [some one may say], let bondage result from *Nature*, as its cause. If you say so, I say 'no;' because that, also, *i.e.*, Nature, also, is dependent on the *conjunction* which is to be mentioned in the next aphorism; because, if it [Nature] were to occasion bondage, even *without* that [conjunction which is next to be mentioned], then bondage would occur even in such cases as the universal dissolution, [when soul is altogether disconnected from the phenomenal].

b. If the reading [in the aphorism] be *nibandhana* [in the 1st case, and not in the 5th], then the construction will be as follows: 'If [you say that] the bondage is caused by Nature,' &c.

c. Therefore, since Nature can be the cause of bondage, only as depending on something else [*i.e.*, on the conjunction to be mentioned in the next aphorism], through this very sort of conjunction [it follows that] the bondage is *reflexional*, like the heat of water due to the conjunction of fire; [water being held to be essentially cold, and to *seem* hot only while the heat continues in conjunction with it].

d. He establishes his own tenet, while engaged on this point, in the very middle [of his criticisms on erroneous notions in regard to the matter; for there are more to come]:

Aph. 19.* [But] not without the conjunction thereof [*i.e.*, of Nature] is there the connexion of that [*i.e.*, of pain] with that [*viz.*, the soul,] which is ever essentially a pure and free intelligence.

a. Therefore, without the conjunction thereof, *i.e.*, without the conjunction of Nature, there is not, to the soul, any connexion with that, *i.e.*, any connexion with bondage; but, moreover, just through that [connexion with Nature] does bondage take place.

b. In order to suggest the fact that the bondage [of the soul] is reflexional [and not inherent in it, either essentially or adventitiously], he makes use of the indirect expression with a double negative, ['not without']. For, if bondage were produced by the conjunction [of the soul] with Nature, as colour is produced by heating [in the case of a jar of black clay, which becomes red in the baking], then, just like that, it would continue even after disjunction therefrom; [as the red colour remains in the jar, after the fire of the brick-kiln has been extinguished, whereas the red colour occasioned in a crystal vase by a China-rose, while it occurs *not without* the China-rose, ceases, on the removal thereof]. Hence, as bondage ceases, on the disjunction [of the soul] from Nature, the bondage is merely reflexional, and neither essential [§ 5. *b*.] nor adventitious [§ 11. *b*.].

c. In order that there may not be such an error as thot of the Vaisheshikas, *viz*., [the opinion that there is] an absolutely real conjunction [of the soul] with pain, he says 'which is ever,' &c. [§ 19]. That is to say: as the connexion of *colour* with essentially pure crystal does not take place without the conjunction of the China-rose [the hue of which, seen athwart the crystal, seems to belong to the crystal], just so the connexion of *pain* with the soul, ever essentially pure, &c., could not take place without the conjunction of some accidental associate; that is to say, pain, &c., cannot arise spontaneously, [any more than a red colour can arise *spontaneously* in the crystal which is essentially pure].

d. This has been declared, in the *Saura*, as follows 'As the pure crystal is regarded, by people, as red, in consequence of the proximity of something [as a China-rose] that lends its colour, in like manner the supreme soul [is regarded as being affected by pain].'

e. In that [aphorism, 19], the perpetual purity means the being ever devoid of merit and demerit; the perpetual intelligence means the consisting of uninterrupted thought; and the perpetual liberatedness means the being ever dissociated from *real* pain: that is to say, the connexion with pain in the shape of a *reflexion* is not a real bondage, [any more than the reflexion of the China-rose is a real stain in the crystal].*

f. And so the maker of the aphorism means, that the cause

of its bondage is just a particular *conjunction* [§ 19. *c*.]. And now enough as to that point.

g. Now he rejects [§ 18. *d*.] certain causes of [the soul's] bondage, preferred by others:

Aph. 20.* Not from Ignorance, too, [does the soul's bondage arise]; because that which is not a reality is not adapted to binding.

a. The word 'too' is used with reference to the previously mentioned 'Time,' &c., [§ 12, which had been rejected, as causes of the bondage, antecedently to the statement, in § 19, of the received cause].

b. Neither, too, does [the soul's] union with bondage result directly from 'Ignorance,' as is the opinion of those who assert non-duality [or the existence of no reality save one (see *Vedanta-sara*, § 20. *b*.)]; because, since their 'Ignorance' is not a real thing, it is not fit to bind; because, that is to say, the binding of any one with a rope merely *dreamt* of was never witnessed.

c. But, if 'Ignorance' *be* a reality [as some assert], then he declares [as follows]:

Aph. 21.* If it ['Ignorance'] *be* [asserted, by you, to be] a reality, then there is an abandonment of the [Vedantic] tenet, [by you who profess to follow the Vedanta].

a. That is to say: and, if you agree that 'Ignorance' *is* a reality, then you abandon your own implied dogma* [see Nyaya Aphorisms I., § 31] of the unreality of Ignorance;' [and so you stultify yourself].

b. He states another objection:

Aph. 22.* And [if you assume 'Ignorance' to be a reality, then] there would be a *duality*, through [there being] something of a different kind [from soul; which you asserters of *non-duality* cannot contemplate allowing].

a. That is to say: if 'Ignorance' is real and without a beginning, then it is eternal, and coordinate with Soul: if [therefore] it be *not* soul, then there is a duality, through [there being] something of a different kind [from soul; and this the Vedantis cannot intend to establish]; because these followers of the *Vedanta*, asserting *non-duality*, hold that there is neither a duality through there being something of the same kind [with soul], nor through there being something of a different kind.

Aph. 23.* If [the Vedanti alleges, regarding 'Ignorance,' that] it is in the shape of both these opposites, [then we shall say 'no,' for the reason to be assigned in the next aphorism].

a. The meaning is: if [the Vedanti says that] 'Ignorance' is not *real*,—else there would be a duality through [there being] something of a different kind [from soul, which a follower of the Vedanta cannot allow],—and, moreover, it is not *unreal*, because we experience its effects; but it is in the shape of something at once real and unreal, [like Plato's Bi êav ìt Di: (see *Vedanta-sara*, § 21)].

Aph. 24.* [To the suggestion that 'Ignorance' is at once real and unreal we say] 'no;' because no such thing is known [as is at once real and unreal.]

a. That is to say: it is not right to say that 'Ignorance' is at once real and unreal. The reason of this he states in the words 'because no such thing,' &c.; because any such thing as is at once real and unreal is not known. For, in the case of a dispute, it is neccssary that there should be an *example* of the thing [*i.e.* (see Nyaya Aphorisms, I., § 25), a case in which all parties are agreed that the property in dispute is really present]; and, as regards *your* opinion, such is not to be found; [for, where is there anything in regard to which both parties are agreed that it is at once real and unreal, as they are agreed that fire is to be met with on the culinary hearth?]: such is the import.

b. Again he ponders a doubt:

Aph. 25.* [Possibly the Vedanti may remonstrate] '*We* are not asserters of any Six Categories, like the *Vaisheshikas* and others.'

a. 'We are not asserters of a definite set of categories [like the *Vaisheshikas*, who arrange all things under six heads, and the *Naiyayikas*, who arrange them under sixteen]. Therefore, we hold that there *is* such a thing, unknown though it be [to people in general], as 'Ignorance' which is at once real and unreal, or [if you prefer it], which differs at once from the real and the unreal [see *Vedanta-sara*, § 21]; because this is established by proofs,' [Scriptural or otherwise, which are satisfactory to *us*, although they may not comply with all the technical requisitions of Gotama's scheme of argumentative exposition (see Nyaya Aphorisms, I., § 35)].

b. By the expression [in the aphorism] 'and others' are meant the *Naiyayikas*; for the *Naiyayika* is an asserter of sixteen categories [see Nyaya Aphorisms, I., § 1].

c. He confutes [this pretence of evading the objection, by disallowing the categories of the Nyaya]:

Aph. 26.* Even although this be not compulsory [that the categories be six, or sixteen], there is no acceptance of the inconsistent; else we come to the level of children, and madmen, and the like.

a. Let there be [accepted] no system of categories [such as that of the *Vaisheshika,* § 25]; still, since *being* and *not-being* are contradictory, it is impossible for disciples to admit, merely on Your Worship's assertion, a thing at once real and unreal, which is inconsistent, contrary to all fitness: otherwise, we might as well accept also the self-contradictory assertions of children and the like: such is the meaning.

b. Certain heretics [deniers of the authority of the Vedas] assert that there exist external objects of momentary duration [individuaily; each being, however, replaced by its facsimile the next instant, so that the uninterrupted series of productions becomes something equivalent to continuous duration], and that by the influence of these the bondage of the soul [is occasioned]. This he objects to, [as follows]:

Aph. 27.* [The bondage] thereof moreover, is not caused by any influence of objects from all eternity.

a. 'Thereof,' *i.e.*, of the soul. An eternal influence of objects, an influence of objects the effect of which, in the shape of a continued stream, has had no commencement,—not by *this*, either, is it possible that the bondage [of the soul] has been occasioned: such is the meaning.

b. He states the reason of this [impossibility]:

Aph. 28.* Also [in my opinion, as well as in yours, apparently], between the external and the internal there is not the relation of influenced and influencer; because there is a local separation; as there is between him that stays at Srughna and him that stays at Pamaliputra.

a. In the opinion of these [persons whose theory we are at present objecting to], the soul is circumscibed, residing entirely within the body; and that which is thus *within* cannot stand in the relation of the influenced and the influencer, as regards an *external* object. Why? Because they are separated in regard to place; like two persons the one of whom remains in Srughna and the other in Pamaliputra: such is the meaning. Because the affection which we call 'influence' (*vasana*) is seen only when there is conjunction, such as that of madder and the cloth [to which it gives its colour], or that of flowers and thc flower-basket [to which they impart their odour.]

b. By the word 'also' the absence or conjunction [between the soul and objects (see § 10)], &c., which he himself holds, is connected [with the matter of the present aphorism].

c. Srughna and Pamaliputra [Palibothra, or Patna] are two several places far apart.

d. But then [these heretics may reply], 'The influence of objects [on the soul] may be asserted, because there is a contact with the object; inasmuch as the soul, according to *us*, goes to the place of the object, just as the senses, according to Your Worship.' Therefore he declares [as follows]:

Aph. 29.* [It is impossible that the soul's bondage should arise] from an influence received in the same place [where the object is; because, in that case], there would be no distinction between the two, [the bond and the free].

a. To complete the sense, we must supply as follows: 'It is impossible that the bondage should arise from an influence received in one and the same place with the object.' Why? Because there would be no distinction between the two, the soul bound and the soul free; because bondage would [in that case] befall the liberated soul, also; [the free soul, according to this hypothesis, being just as likely to come across objects as any other]: such is the meaning.

b. Here he ponders a doubt:

Aph. 30.* If [the heretic, wishing to save his theory suggests that a difference between the two cases (see § 29) *does* exist] in virtue of the *unseen*, [*i.e.*, of merit and demerit, then he will find his answer in the next aphorism].

a. That is to say, [the heretic may argue]: 'But then, granting that they [the free soul and the bound] are alike in respect of their coming into contact with objects, when they become conjoined with them in one and the same locality; yet the *reception of the influence* may result merely from the force of the *unseen*, [*i.e.*, from the merit and demerit of this or that soul; the soul that is liberated alike from merit and demerit being able to encounter, with impunity, the object that would enchain one differently circumstanced]': if [*this* be urged, then we look forward].

b. This he disputes, [as follows]:

Aph. 31.* They cannot stand in the relation of deserver and bestower, since the two do not belong to one and the same time.

a. Since, in thy opinion, the agent and the patient are distinct, and do not belong to the same time [believing, as thou heretically dost, not only that *objects* (see § 26. *b*.) momentarily perish and are replaced, but that the duration of *souls*, also, is of a like description], there is positively no such relation [between the soul at one time and its successor at another] as that of deserver and bestower [or transmitter of its merits or demerits]; because it is impossible that there should be an influence of objects [§ 27] taking effect on a patient [say, the soul of to-day], occasioned by the 'unseen' [merit or demerit] belonging to an agent [say, the soul of yesterday, which, on the hypothesis in question, is a numerically different individual]: such is the meaning.

b. He ponders a doubt:

Aph. 32.* If [the heretic suggests that] the case is like that of the ceremonies in regard to a son, [then he will find his reply by looking forward].

a. But then [the heretic, admitting the principle that the merit or demerit of an act belongs entirely to the agent, may urge that], as the son is benefited by ceremonies in regard to a son, such as that [ceremony (see Colebrooke's 'Hindu Law,' Vol. III.) celebrated] in anticipation of conception, which [no doubt] belongs to the *father* [who performs the ceremonies, to propitiatc the gods], in like manner there may be an influence of objects on the experiencer [say, the soul of to-day], through the 'unseen' [merit or demerit] that belongs even to a different subject [say, the soul of yesterday]: such is the meaning [of the heretic].

b. He refutes this, by showing that the illustration is not a fact:

Aph. 33.* [Your illustration proves nothing;] for, in that case, there is no one permanent soul which could be consecrated by the ceremonies in anticipation of conception, &c.

a. 'In that case,' *i.e.*, on thy theory, too, the benefit of the son, by [means of the performance of] the ceremonies in anticipation of conception, &c., could not take place; 'for,' *i.e.*, because, on that theory, there is not one [self-identical] soul, continuing from the [time of] conception to birth, which could be consecrated [by the ceremonies in question], so as to be a fit subject for the duties that pertain to the time subsequent to birth [such as the investiture with the sacred thread, for which the young Brahman would not be a fit subject, if the ceremonies in anticipation of his conception had been omitted]: and thus your illustration is not a real one, [on your *own* theory: it is not a thing that you can assert as a fact].

b. And, according to *my* theory, also, your illustration is not a fact; seeing that it is possible that the benefit to the son should arise from the 'unseen' [merit] deposited in the son by means of the ceremony regarding the son: for it is an implied tenet [of my school], that it [the soul] is permanent [in its self-identity]; and there is the injunction [of Manu, (Ch. II., v. 26), with regard to the ceremonies in question, which proceeds on the same grounds].

c. Some other heretic may encounter us, on the strength of [the argument here next stated, *viz.*,] 'But then, since *bondage*, also, [like everything else] is momentary, let this bondage have nothing determinate for its cause, or *nothing at all* for its cause,' [which view of matters is propounded in the next aphorism]:

Aph. 34.* Since there is no such thing as a permanent result [on the heretical view], the momentariness [of bondage, also, is to be admitted].

a. 'Of bondage': this must be supplied, [to complete the aphorism].

b. And thus the point relied on is, that it [*i.e.*, bondage] have no cause at all. And so this is the application [of the argument, *viz.*]:

(1) Bondage, &c., is momentary;

(2) Because it exists,

(3) [Everything that exists is momentary,] as the apex of the lamp-flame, or the like.

c. And [continues the heretic,] this [reason, *viz.*, 'existence'] does not extend *unduly* [as you may object,] to the case of a jar, or the like; because *that*, also [in my opinion], is like the subject in dispute; [in being momentary]. This [in fact] is precisely what is asserted in the expression, 'since there is *no such thing* as a permanent result' [§ 34].

d. He objects [to this heretical view]:

Aph. 35.* No, [things are *not* momentary in their duration]; for the absurdity of this is proved by *recognition*.

a. That is to say: nothing is momentary; because the absurdity of its being momentary follows from the opposite argument [to that under § 34. *b*.]. taken from such facts of recognition as, 'what I saw, that same do I touch,' [an argument which may be stated as follows], *viz.*:

(1) Bondage, &c., is permanent;

(2) Because it exists,

(3) [Everything that exists is permanent,] as a jar, or the like.

Aph. 36.* And [things are not momentary;] because this is contradicted by Scripture and reasoning.

a. That is to say: nothing is momentary; because the general principle, that the whole world, consisting of effects and causes, is momentary, is contradicted by such texts as this, *viz.*, '[All] this, O ingenuous one, was antecedently existing,' and by such Scriptural and other arguments as this, *viz.*, 'How should what exists proceed from the non-existent?'

Aph. 37.* And [we reject the argument of this heretic;] because his instance is not a fact.

a. That is to say: the general principle of the momentariness [of all things] is denied; because this momentary churacter does not [in fact] belong to the apex of the lamp-flame, &c., the instance [on which thou, heretic, dost ground thy generalization, (§ 34. *b.*)]. Moreover, thou quite errest in regard to momentariness, in that instancc, from not taking account of the minute and numerous instants [really included in a duration which seems to thee momentary]: such is the import.

b. Moreover, if the momentary duration, &c., [of things] be asserted, then there can be no such thing as the relation of cause and effect, in the case of the earth and the jar, and the like. And you must not say that there *is* no such thing as that [relation of cause and effect]; because it is proved to be a reality by the fact that, otherwise, there would be no such thing as the efforts of him who desires an effect; [and who, therefore, sets in operation the causes adapted to its production]. With reference to this, he declares [as follows]:

Aph. 38.* It is not between two things coming simultaneously into existence, that the relation of cause and effect exists.

a. Let us ask, does the relation of product and [material] cause exist between the earth and the jar, as *simultaneously* coming into [their supposed momentary] existence, or as successive? Not the first; because there is nothing to lead to such an inference, and because we should not [in that case] find the man, who wants a jar, operating with earth, &c., [with a view to the jar's *subsequent* production]. Nelther is it the last; in regard to which he declares [as follows]:

Aph. 39.* Because, when the antecedent departs the consequent is unfit [to arise, and survive it].

a. The relation of cause and effect is, further, inconsistent with the theory of the momentary duration of things; because, at the time when the antecedent, *i.e.*, the cause, departs, the consequent, *i.e.*, the product, is 'unfit,' *i.e.*, is not competent to arise; because, that is to say, a product is cognized only by its inhering in [and being substantially identical with, however formally different from,] its substantial cause, [and is incapable, therefore, of surviving it].

b. With reference to this same [topic, *viz.*, the] substantial cause, he mentions another [the converse] objection [to the theory of the momentary duration of things]:

Aph. 40.* Moreover, not [on the theory of the momentary duration of things can there be such a relation as that of cause and effect]; because, while the one [the antecedent] exists, the other [the consequent] is incompatible, because the two keep always asunder.

a. To complete [the aphorism], we must say, 'moreover, [on the theory objected to], there can be no such relation as that of cause and effect; because, at the time when the antecedent exists, the consequent cannot coexist with it, the two being mutually exclusive.' The two suggesters of the relation of cause and effect, in product and substance, are (1) concomitancy of affirmatives, that, while the product exists, the substance thereof exists, and (2) this concomitancy of negatives, that, when the substance no longer exists, the product no longer exists: and these two [conditions, on *your* theory] cannot be; because, since things [in your opinion,] are momentary in their duration, the two [*viz.*, the substance and the product], inasmuch as they are antecedent and consequent, belong to opposite times, [and cannot, therefore, coexist; for the product, according to you, does not come into existence until its substance has perished, which is contrary to the nature of the causal relation just defined].

b. But then, [the heretic may say, do not let the *coexistence* of substance and product be insisted upon, as indispensable to the causal relation between the two, but] 'let the nature of a cause belong to the substantial cause, as it belongs to the *instrumental* cause, in respect merely of its *antecedence.*' To this he replies:

Aph. 41.* If there were merely *antecedence*, then there would be no determination [of a substantial or material cause, as distinguished from an instrumental cause].

a. And it could not be determined that this was the *substance* [of this or that product], on the granting of nothing more than its *antecedence* [to the product]; because antecedence constitutes no distinction between it and the *instrumental* causes; for, [as we need scarcely remind you], that there *is* a distinction between instrumental and substantial causes, the whole world is agreed: such is the meaning.

b. Other heretics say: 'Since nothing [really] exists, except *Thought*, neither does *Bondage*; just as the things of a dream [have no real existence]. Therefore it has *no* cause; nor it is absolutely *false.*' He rejects the opinion of these [heretics]:

Aph. 42.* Not Thought alone exists; because there is the intuition of the external.

a. That is to say: the *reality* is not *Thought* alone; because external objects, also, are proved to exist, just as Thought is, by intuition.

b. But then [these heretics may rejoin], 'From the example of intuitive perception in *dreams* [see Butler's 'Analogy,' Part I., Ch. I.], we find this [your supposed evidence of objective reality] to exist, even in the *absence* of objects!' To this he replies:

Aph. 43.* Then, since, if the one does not exist, the other does not exist, there is a void, [*i.e.*, nothing exists at all].

a. That is to say: if external things do not exist, then a mere void offers itself. Why? Because, if the external does not exist, then *thought* does not exist; for it is *intuition* that proves the objective: and, if the intuition of the external did not establish the objective, then the intuition of *thought*, also, would not establish [the existence of] thought.

b. 'Then *let* the reality be a mere void; and, therefore, the searching for the cause of Bondage is unfitting, *just because a void is all*:' with such a proposal [as recorded in the next aphorism] does [some one who may claim the title of] the very crest-gem of the heretics rise up in opposition:

Aph. 44.* The reality is a void: what is perishes; because to perish is the habit of things.

a. The void alone [says this prince of heretics, or the fact that nothing exists at all] is the reality, [or the only truth]. Since everything that exists perishes, and that which is perishable is false, as is a dream, therefore, as of all things the beginnings and endings are merely nonentities, Bondage, &c., in the midst [of any beginning and ending], has merely a momentary existence,—is phenomenal, and not real. Therefore, *who* can be bound by *what*? This [question] is what we rest upon. The reason assigned for the perishableness of whatever exists is, 'because to perish is the habit of things;' because to perish is the *very nature* of things: but nothing continues, after quitting its own *nature*; [so that nothing could continue, if it *ceased* to perish]: such is the meaning.

b. He rejects [this heretical view]

Aph. 45.* This is a mere counter-assertion of unintelligent persons.

a. 'Of unintelligent persons,' *i.e.*, of blockheads, this is 'a mere counter-assertion,' *i.e.*, a mere *idle* counter-assertion that a thing must needs be perishable, *because it exists*; [and such an assertion is idle,] because things that are not made up of parts, since there is no cause of the destruction of such things, cannot perish.

b. [But] what need of many words? It is not the fact, that even *products* perish; [for] just as, by the cognition that 'the jar is old' [we mean that it has passed from the condition of new to that of old], so, too, by such a cognition as this, that 'the jar has passed away,' it is settled only that the jar, or the like, *is in the condition* of having passed away.

c. He states another objection [to the heretical view]:

Aph. 46.* Moreover, this [nihilistic theory is not a right one]; because it has the same fortune as both the views [which were confuted just before].

a. This view, moreover [§ 44], is not a good one; because it has the same fortune as, *i.e.*, is open to similar reasons for rejection as, the theory that external things are momentary [§ 26. *b.*], and as the theory that nothing exists besides Thought [§ 41. *b.*]. The reason for the rejection of the theory that things are momentary in their duration. *viz.* [as stated in § 35], the fact of *recognition*, &c., [which is, at least, as little consistent with Nihilism as it is with the momentary duration of things], and the reason for the rejection of the theory that nothing exists besides Thought. *viz.* [as stated in § 42], the intuition of the external, &c., apply equally here [in the case of Nihilism]: such is the import.

b. Moreover, as for the opinion which is accepted by these [heretics], *viz.*, '*Let the mere void* [of absolute nonentity] be the soul's aim [and *summum bonum*], since herein consist at once the cessation of pain [which cannot continue, when there is absolutely *nothing*], and also the means thereof [since there can be no further means required for the removal of anything, if it be settled that the thing positively does not exist],' this too, can hardly be: so he declares [as follows]:

Aph. 47.* In neither way [whether as a means, or as an end,] is this [annihilation] the soul's aim.

a. 'Let the void [of mere nonentity] be the soul's aim, whether as consisting in the cessation of pain, or as presenting the means for the cessation of pain,' [says the heretic. And this cannot be; because the [whole] world agrees, that the aim of the soul consists in the joys, &c., that shall abide *in it*; that is to say, because [*they* hold, while] *you* do not hold, that there is a *permanent* soul, [(see § 33) in respect of which the liberation or beatification would be possible, or even predicable].

b. Now [certain] other things, also, entertained, as causes of [the soul's] bondage, by [imperfectly instructed] believers, remaining over and above those [proposed by unbelievers, and] already rejected, are to be set aside:

Aph. 48.* Not from any kind of motion [such as its entrance into a body, does the soul's bondage result].

a. 'Bondage' [required to complete the aphorism] is understood from the topic [of discussion].

b. The meaning is, that the soul's bondage, moreover, does not result from any sort of *motion*, in the shape, for instance, of its entrance into a body.

c. He states a reason for this:

Aph. 49.* Because this is impossible for what is inactive, [or in other words without motion]

a. That is to say: because this is impossible, *i.e.*, *motion* is impossible, in the case of the soul, which is inactive, [because] all-pervading, [and, therefore, incapable of changing its place].

b. But then [the objector may say], 'Since, in the books of Scripture and of law, we hear of its *going* and *coming* into this world and the other world, let soul be [not all-pervading, as you allege, but] merely limited [in its extent]: and to this effect, also, is the text, 'Of the size of the thumb is the soul, the inner spirit,' and the like: [but] this conjecture he repels:

Aph. 50.* [We cannot admit that the soul is other than all-pervading; because] by its being limited, since it would come under the same conditions as jars, &c., there would be a contradiction to our tenet [of its imperishableness].

a. That is to say: and, if the soul were admitted to be, like a jar, or the like, limited, *i.e.* circumscribed [in dimension], then, since it would resemble a jar, or the like, in being made up of parts, and [hence] in being perishable, &c., this would be contrary to our settled principle, [that the soul is imperishable].*

b. He now justifies the text [see § 49. *b.*] referring to the *motion* [of the soul, by showing that the motion is not really of the soul, but of an accessory]:

Aph. 51.* The text regarding the *motion* [of the soul], moreover, is [applicable, only] because of the junction of an *attendant*; as in the case of the Ether [or *Space*, which moves not, though we talk of the space enclosed in a jar, as moving with the jar].

a. Since there are such proofs of the soul's unlimitedness, as the declaration that 'It is eternal, omnipresent, permanent,' the text regarding its motion is to be explained as having reference to a movement pertaining [not to the soul, but] to an attendant; for there is the text, 'As the Ether [or space] included in a jar, when the jar is removed, [in this case] the *jar* may be removed, but not the space; and in like manner is the soul, which is like the sky, [incapable of being moved]'; and because we may conclude that the motion [erroneously supposed to belong to the soul (49. *b*.),] belongs to *Nature* [see *Vedanta Aphorisms*, Part I., § 4. *l*.], from such maxims as this, that 'Nature does the works the fruits of which are blissful or baneful; and it is wilful *Nature* that] in the three worlds, reaps these': such is the import.

b. It has already been denied [§ 16] that the bondage [of the soul] is occasioned by works] in the shape either of enjoined or of forbidden actions. Now he declares that the bondage, moreover] does not arise from the 'unseen' [merit or demerit] resulting therefrom:

Aph. 52.* Nor, moreover, [does the bondage of the soul result from the merit or demerit arising] from works; because these belong not thereto.

a. That is to say: the bondage of the soul does not arise directly from the 'unseen' [merit or demerit] occasioned by works. Why? Because this is no property thereof, *i.e.*, because this [merit or demerit (see § 16. *a*.)] is no property of the soul.

b. But then [some one may say], '*Let* it be that the bondage resulting from the 'unseen,' *i.e.*, the merit [or demerit] even of another, should attach to a different person;' whereupon he declares [as follows]:

Aph. 53.*4 If the case were otherwise [than as I say], then it [the bondage of the soul might extend unduly, even to the emancipated].

a. That is to say: if the case were otherwise, if bondage and its cause were under other conditions [than we have declared them to be], then there might be an undue extension; bondage would befall even the emancipated, [for the same reasons as those stated under § 16. *a*.].

b. What need of so much [prolixity]? He states a general objection why the bondage of soul cannot result from any one or other [of these causes], beginning with its essence [see § 6. *b*.], and ending with its [supposed] works [see § 16]; inasmuch as it is contrary to Scripture, [that any one of these should be the cause]:

Aph. 54.* And this [opinion, that the bondage of the soul arises from any of causes alleged by the heretics,] is contrary to such texts as the one that declares it [the soul] to be without qualities: and so much for that point.

a. And, if the bondage of the soul arose from any one or other of those [supposed causes already treated of,] among which its essential character [§ 6. *b*.] is the first, this would be contradictory to such texts as, 'Witness, intelligent, alone, and without the [three] qualities [is the soul]:' such is the meaning.

b. The expression 'and so much for that point' means, that the investigation of the cause of the bondage [of the soul] here closes.

c. The case, then, stands thus: since [all] other [theories] are overthrown by the declaratory aphorisms, 'There would be no fitness in the enjoining' [see § 7], &c., it is ascertained that the immediate cause of the bondage [of the soul] is just the conjunction of Nature and of the soul.

d. But then, in that case, [some one may say], this conjunction of Nature and of the soul [§ 54. *c*.], whether it be essential, or adventitiously caused by Time or something else [§ 5. *b*.], must occasion the bondage even of the *emancipated*. Having pondered this doubt, he disposes of it [as follows]:

Aph. 55.* Moreover, the conjunction thereof does not, through non-discrimination, take place [in the case of the emancipated]; nor is there a parity, [in this respect, between the emancipated and the unemancipated].

a. 'The conjunction thereof,' *i.e.*, the conjunction of Nature and of the soul; this conjunction, moreover, does not take place again 'through non-discrimination,' *i.e.*, through the want of a discrimination [between Nature and soul] in the emancipated, [who *do* discriminate, and who thus avoid the conjunction which others, failing to discriminate, incur, and thus fall into bondage]: such is the meaning. And thus the emancipated and the bound are *not* on a level, [under the circumstances stated at § 54. *c.*]: such is the import.

Aph. 56.* Bondage arises from the error [of not discriminating between Nature and soul].

a. Having thus declared the cause of that [bondage] which is to be got rid of, he declares the means of getting rid of it:]

Aph. 56.* The removal of it is to be effected by the necessary means, just like darkness.

a. The necessary means, established throughout the world, in such cases as 'shell-silver' [*i.e.*, a pearl-oyster-shell mistaken for silver], *viz.*, the *immediacy* of discrimination, by *this* alone is 'its removal,' *i.e.*, the removal of the non-discrimination [between Nature and soul], to be effected, and not by *works*, or the like: such is the meaning: just as darkness, the dark, is removed by light alone, [and by no other means].

b. 'But then [some one may say], if merely the non-discrimination of Nature and soul be, through the conjunction [of the two, consequent on the want of discrimination], the cause of bondage, and if merely the discrimination of the two be the cause of liberation, then there would be liberation, even while there remained the conceit of [one's possessing] a body, &c.; and this is contrary to Scripture, to the institutes of law, and to sound reasoning.' To this he replies:

Aph. 57.* Since the non-discrimination of other things [from soul] results from the non-discrimination of *Nature* [from soul], the cessation of this will take place, on the cessation of that [from which it results].

a. By reason of the non-discrimination of *Nature* from the soul, what non-discrimination of *other* things there *is*, such as the non-discrimination of the *understanding* [as something other than the soul], *this* necessarily ceases, on the cessation of the non-discrimination of Nature; because, when the non-discrimination of the understanding, for example, [as something other than soul,] does occur, it is *based* on the non-discrimination [from soul] of that cause to which there is none antecedent [*viz.*, Nature]; since the non-discrimination of an *effect* [and the 'understanding' is an effect or product of Nature,] is, itself, an effect, [and will, of course, cease, with the cessation of its cause].

b. The state of the case is this: as, when the soul has been discriminated from the *body*, it is impossible but that it should be discriminated from the *colour* and other [properties], the effects of the body, [which is the substantial cause of its own properties]; so, by parity of reasoning, from the departure of the cause, when soul, in its character of *unalterableness*, &c., has been discriminated from *Nature*, it is impossible that there should remain a conceit of [the soul's being any of] the *products* thereof [*i.e.*, of Nature], such as the 'understanding,' and the like, which have the character of being *modifications* [of primal Nature, while the soul, on the other hand, is a thing unalterable].*

c. But then [some one may say], 'What proof is there that there is a conceit [entertained by people in general,] of a *Nature* [or primal principle] different from the conceit of an 'understanding,' &c., [which, you tell us, are products of this supposed first principle]? For all the various conceits [that the soul falls into], such as, 'I am ignorant,' and so on, can be accounted for on the ground simply of an 'understanding,' &c., [without postulating a primal Nature which is to assume the shape of an 'understanding,' &c.]:' well, if any one says this, I reply, 'no;' because, unless there were such a thing as Nature, we could not account for such conceits as the following, *viz.*, 'Having died, having died, again, when there is a creation, let me be a denizen of Paradise, and not of hell;' because no *products*, such as the 'understanding,' when they have perished, can be created anew, [any more than a gold-bracelet, melted down, can be reproduced, though another like it may be produced from the materials].

d. Moreover, it is inadmissible to say that men's conceit of [the identity of themselves with their] 'understanding,' &c., is [the *primary* cause of the soul's bondage, and is] not preceded by anything; because 'understanding' and the rest [as you will not deny] are *effects*. Now, while it is to be expected that there should be some predetermining agency to establish a conceit of [ownership in, or of one's identity with,] any *effects*, it is clear that it is a conceit of [ownership, &c.,] in respect of the *cause*, and nothing else, that must be the predetermining agency: for we see this in ordinary life; and our theories are bound to conform [deferentially] to experience. For [to explain,] we see, in ordinary life, that the conceit of [the ownership of] the grain, &c., produced by a field, results from the conceit of [the ownership of] the field; and, from the conceit of [the ownership of] gold, the conceit of [the ownership of] the bracelets, or other things, formed of that gold; and, by the removal of these [*i.e.*, the removal of the logically antecedent conceits, that the field, or the gold, is one's property], there is the removal of those, [*i.e.*, the removal of the conceits that the grain, &c., and that the bracelets, &c., the corresponding products or effects of the field and of the gold, are one's property: and so the soul will cease to confound itself with the 'understanding,' when it ceases to confound itself with Nature, of which the 'understanding' is held to be a product].

e. [And, if it be supposed that we thus lay ourselves open to the charge of a *regressus ad infinitum*, seeing that, whatever we may assign as the *first* cause, we may, on our own principles, be asked what was the 'predetermining agency' in regard to *it*; or if it be supposed thut we are chargeable with reasoning in a circle, when we hold that the soul's confounding itself with Nature is the cause of its *continuing* so to confound itself, and its continuing so to confound itself is, reciprocally, the cause why it confounds itself; we reply, that] there is no occasion to look for any other 'predetermining agency,' in the case of the conceit of [the identity of the soul with] Nature, or in the case of the self-continuance thereof, [*i.e.*, of that error of confounding one's self with Nature]; because [these two are alike] without antecedent, like seed and sprout, [of which it is needless to ask which is the first; the old puzzle, 'which was first, the acorn, or the oak?' being a frivolous question].

f. But then [some one may say], if we admit the soul's bondage [at one time], and its freedom [at another], and its discrimination [at one time], and its non-discriminatian [at another], then this is in contradiction to the assertion [in § 19], that it is 'ever essentially a pure and free intelligence;' and it is in contradiction to such texts as this, *viz.*, 'The absolute truth is this, that neither is there destruction [of the soul], nor production [of it]; nor is it bound, nor is it an effecter [of any work], nor is it desirous of liberation, nor is it, indeed, *liberated*; [seeing that that cannot desire or obtain liberation, which was never *bound*].' This [charge of inconsistency] he repels:

Aph. 58.* It is merely verbal, and not a reality [this so-called bondage of the soul]; since it [the bondage] *resides* in the *mind*, [and not in the soul].

a. That is to say: since bondage, &c., all reside only in the *mind* [and not in the soul], all this, as regards the soul, is merely verbal, *i.e.*, it is *vox et praeterea nihil*; because is is merely a *reflexion*, like the redness of [pellucid] crystal [when a China-rose is near it], but not a reality, with no false imputation, like the redness of the China-rose itself. Hence there is *no* contradiction to what had been said before, [as the objector (under § 57. *f.*) would insinuate]: such is the state of the case.

b. But then, if bondage, &c., as regards the soul, be merely verbal, let them be set aside by *hearing* [that they are merely verbal], or by argument [establishing that they are so]. Why, in the Scripture and the Law, is there enjoined, as the cause of liberation, a discriminative knowledge [of Soul, as distinguished from Non-soul], going the length of *immediate cognition*? To this he replies:

Aph. 59.* Moreover, it [the non-discrimination of Soul from Nature,] is not to be removed by argument; as that of a person perplexed about the points of the compass [is not to be removed] without immediate cognition.

a. By 'argument' we mean thinking. The word 'moreover' is intended to aggregate [or take in, along with 'argument'] 'testimony,' [or verbal authority, which, no more than 'argument,' or inference, can remove the evil, which can be removed by nothing short of direct intuitive *perception* of the real state of the case].

b. That is to say: the bondage, &c., of the soul though [granted to be] merely verbal, are not to be removed by merely hearing, by inferring, without immediate cognition, without directly perceiving; just as the contrariety in regard to the [proper] direction, though merely verbal [as resulting from misdirection], in the case of a person who is mistaken as to the points of the compass [and hence as to his own bearings], is not removed by testimony, or by inference, without immediate cognition, *i.e.*, without [his] directly perceiving [how the points of the compass really lie, to which immediate perception 'testimony,' or 'inference,' may conduce, but the necessity of which these *media*, or instruments of knowledge; cannot supersede].

c. Or it [Aph. 59] may be explained as follows, *viz.*: But then, [seeing that] it is declared, by the assertion [in Aph. 56], *viz.*, that 'The removal of it is to be effected by the necessary means,' that knowledge, in the shape of discrimination [between soul and Nature], is the remover of *non*-discrimination [in regard to the matter in question], tell us, is that knowledge of a like nature with the hearing [of Testimony], &c.? Or is it something peculiar? A reply to this being looked for, he enounces the aphorism [§ 59]: 'Moreover, it is not to be removed by argument,' &c. That is to say: non-discrimination is not excluded, is not cut off, by argument, or by testimony, unless there be discrimination as an immediate perception; just as is the case with one who is bewildered in regard to [his] direction; because the only thing to remove an *immediate* error is an immediate individual perception [of the truth. For example, a man with the jaundice perceives *white* objects as if they were *yellow*. He may *infer* that the piece of chalk which he looks at is really white; or he may believe the *testimony* of a friend, that it *is* white; but still nothing will remove his erroneous *perception* of yellowness in the chalk, except a direct perception of whiteness.

d. Having thus, then, set forth the fact that Liberation results from the immediate discrimination [of Soul from Nature], the next thing to be set forth is the 'discrimination' [here referred to].

e. This being the topic, in the first place, since only if Soul and Nature exist, liberation can result from the discrimination of the one from the other, therefore that 'instrument of right knowledge' (*pramana*) which establishes the existence of these [two *imperceptible* realities] is [first] to be set forth:

Aph. 60.* The knowledge of things imperceptible is by means of Inference; as that of fire [when not directly perceptible,] is by means of smoke, &c.

a. That is to say: 'of things imperceptible,' *i.e.*, of things not cognizable by the senses, *e.g.*, Nature and the Soul, 'the knowledge,' *i.e.*, the fruit lodged in the soul, is brought about by means of that instrument of right knowledge [which may be called] 'Inference' (*anumana*), [but which (see Nyaya Aphorisms, I., § 5) is, more correctly, 'the recognition of a Sign']; as [the knowledge that there is] fire [in such and such a locality, where we cannot directly perceive it,] is brought about by the 'recognition of a Sign,' occasioned by smoke, &c.

b. Moreover, it is to be understood that that which is [true, but yet is] not established by 'Inference,' is estalished by Revelation. But, since 'Inference' is the chief [among the instruments of knowledge], in this [the Samkhya] System, 'Inference' only is laid down [in the aphorism,] as the *chief* thing; but Revelation is not disregarded [in the Samkhya system; as will be seen from Aph. 88 of this Book].

c. He [next] exbibits the order of creation of those things among which Nature is the first, and the relation of cause and effect [among these, severally], preparatorily to the argument that will be [afterwards] stated:

Aph. 61.* Nature (*prakriti*) is the state of equipoise of Goodness (*sattwa*) Passion (*rajas*), and Darkness (*tamas*): from Nature [proceeds] Mind (*mahat*); from Mind, Self-consciousness (*ahankara*); from Self-consciousness, the five Subtile Elements (*tan-matra*), and both sets [external and internal,] of Organs (*indriya*); and, from the Subtile Elements, the Gross Elements (*sthula-bhuta*). [Then there is] Soul (*purusha*). Such is the class of twenty-five.

a. 'The state of equipoise' of the [three] things called 'Goodness,' &c., is their being neither less nor more [one than another]; that is to say, the state of *not* being [developed into] an *effect* [in which one or other of them predominates]. And thus 'Nature' is the triad of 'Qualities' (*guna*), distinct from the products [to which this triad gives rise]: such is the complete meaning.

b. These things, *viz.*, 'Goodness,' &c., [though spoken of as the three Qualities], are not 'Qualities' (*guna*) in the *Vaisheshika* sense of the word; because [the 'Qualities' of the *Vaisheshika* system have, themselves, *no* qualities (see *Kanada*'s 16th Aph.); while] *these* have the qualities of Conjunction, Disjunction, Lightness, Force, Weight, &c. In this [Samkhya] system, and in Scripture, &c., the word 'Quality' (*guna*) is employed [as the name of the three things in question], because they are subservient to Soul [and, therefore, hold a secondary rank in the scale of being], and because they form the *cords* [which the word *guna* also signifies], *viz.*, 'Mind,' &c., which consist of the three [so-called] 'Qualities,' and which *bind*, as a [cow, or other] brute-beast, the Soul.

c. Of this [Nature] the principle called 'the great one' (*mahat*), *viz.*, the principle of Understanding, (*buddhi*), is the product. 'Self-consciousness' is a conceit [of separate personality]. Of this there are two products, (1) the 'Subtile Elements' and (2) the two sets of 'Organs.' The 'Subtile Elements' are [those of] Sound, Touch, Colour, Taste, and Smell. The two sets of 'Organs,' through their division into the external and the internal, are of eleven kinds. The products of the 'Subtile Elements' are the five 'Gross Elements.' But 'Soul' is something distinct from either product or cause. Such is the class of twenty-five, the aggregate of things. That is to say, besides these there is nothing.

d. He [next], in [several] aphorisms, declares the order of the inferring [of the existence of these principles, the one from the other]:

Aph. 62.* [The knowledge of the existence] of the five 'Subtile Elements' is [by inference,] from the 'Gross Elements.'

a. 'The knowledge, by inference,' so much is supplied, [to complete the aphorism, from Aph. 60].

b. Earth, &c., the 'Gross Elements,' are proved to exist, by Perception; [and] thereby [*i.e.*, from that Perception; for Perception must precede Inference, as stated in Gotama's 5th Aphorism,] are the 'Subtile Elements' inferred, [the Empedocles]. And so the application [of the process of inference to the case] is as follows:

(1) The Gross Elements, or those which have not reached the absolute limit [of simplification, or of the atomic], consist of things [Subtile Elements, or Atoms,] which have distinct qualities; [the earthy element having the distinctive quality of Odour; and so of the others]:

(2) Because they are gross;

(3) [And everything that is gross is formed of something less gross, or, in other words, more subtile,] as jars, webs, &c.; [the gross web being formed of the less gross threads; and so of the others].

Aph. 63.* [The knowledge of the existence] of Self-consciousness is [by inference,] from the external and internal [organs], and from these ['Subtile Elements,' mentioned in Aph. 62].

a. By inference from [the existence of] the external and internal organs, and from [that of] these 'Subtile Elements,' there is the knowledge of [the existence of such a principle as] Self-consciousness.

b. The application [of the process of inference to the case] is in the following [somewhat circular] manner:

(1) The Subtile Elements and the Organs are made up of things consisting of Self-consciousness:

(2) Because they are products of Self-consciousness:

(3) Whatever is not so [*i.e.*, whatever is *not* made out of Self-consciousness] is not thus [*i.e.*, is not a *product* of Self-

consciousness]; as the Soul, [which, not being made up thereof, is not a product of it].

c. But then, if it be thus [*i.e.*, if it be, as the Sankhyas declare, that all objects, such as jars, are made up of Self-consciousness, while Self-consciousness depends on Understanding,' or 'Intellect,' or 'Mind,' the *first* product of Nature' (see Aph. 61)], then [some may object, that], since it would be the case that the Self-consciousness of the potter is the material of the jar, the jar made by him would disappear, on the beatification of the potter, whose internal organ [or 'Understanding'] then surceases. And this [the objector may go on to say,] is not the case; because *another* man [*after* the beatification of the potter,] recognizes that 'This is that same jar [which, you may remember, was fabricated by our deceased acquaintance].'

d. [In reply to this we say,] it is *not* thus; because, on one's beatification, there is an end of only those modifications of his internal organ [or 'Intellect'] which could be causes [as the *jar* no longer can be,] of the emancipated soul's *experiencing* [either good or ill], but not an end of the modifications of intellect in general, nor [an end] of intellect altogether: [so that we might spare ourselves the trouble of further argument, so far as concerns the objection grounded on the assumption that the intellect of the potter *surceases*, on his beatification: but we may go further, and admit, for the sake of argument, the surcease of the 'intellect' of the beatified potter, without conceding any necessity for the surcease of his pottery. This alternative theory of the case may be stated as follows]:

e. Or [as Berkeley suggests, in his Principles of Human knowledge, Ch. vi.], let the Self-consciousness of the *Deity* be the cause why jars and the like [continue to exist], and not the Self-consciousness of the potter, &c., [who may lose their Self-consciousness, whereas the Deity, the sum of all life, *Hiranyagarbha* (see *Vedanta-shara*, § 62), never loses *his* Self-consciousness, while aught living continues].

Aph. 64.* [The knowledge of the existence] of Intellect is [by inference,] from that [Self-consciousness, § 63].

a. That is to say: by inference from [the existence of] 'that,' *viz.*, Self-consciousness, which is a product, there comes the knowledge of 'Intellect' (*buddhi*), the *great* 'inner organ' (*anta-karana*), [hence] called 'the great one' (*mahat*), [the existence of which is recognized] under the character of the *cause* of this [product, *viz.*, Self-consciousness].

b. And so the application [again rather circular, of the process of inference to the case,] is as follows:

(1) The thing called Self-consciousness is made out of the things that consist of the moods of judgment [or mind];

(2) Because it is a thing which is a product of judgment [proceeding in the Cartesian order of *cogito, ergo sum*; and]

(3) Whatever is not so [*i.e.*, whatever is *not* made out of judgment, or mental assurance], is not thus [*i.e.*, is not a product of mental assurance]; as the Soul, [which is not made out of this or of anything antecedent], &c.

c. Here the following reasoning is to be understood: Every one, having first determined anything under a concept [*i.e.*, under such a form of thought as is expressed by a general term; for example, that this which presents itself is a jar, or a human body, or a possible action of one kind or other], after that makes the judgment, 'This is I,' or 'This ought to be done by me,' and so forth: so much is quite settled; [and there is no dispute that the fact is as here stated]. Now, having, in the present instance, to look for some *cause* of the thing called 'Self-consciousness' [which manifests itself in the various judgments just referred to], since the relation of cause and effect subsists between the two functions [the occasional conception, and the subsequent occasional judgment, which is a function of Self-consciousness], it is assumed, for simplicity, merely that the relation of cause and effect exists between the two substrata to which the [two sets of] functions belong; [and this is sufficient,] because it follows, as a matter of course, that the occurrence of a *function* of the effect must result from the occurrence of a *function* of the cause; [nothing, according to the Samkhya, being in any product, except so far, and in such wise, as it preexisted in the cause of that product].

Aph. 65.* [The knowledge of the existence] of Nature is [by inference,] from that ['Intellect,' § 64].

a. By inference from [the existence of] 'that,' *viz.*, the principle [of Intellect, termed], 'the Great one,' which is a *product*, there comes the knowledge of [the existence of] Nature, as [its] cause.

b. The application [of the process of inference to the case] is as follows:

(1) Intellect, the affections whereof are Pleasure, Pain, and Dulness, is produced from something which has these affections, [those of] Pleasure, Pain, and Dulness:

(2) Because, whilst it is a *product* [and must, therefore, have arisen from something consisting of that which itself now consists of], it consists of Pleasure, Pain, and Dulness; [and]

(3) [Every *product* that has the affections of, or that occasions, Pleasure, Pain, or Dulness, takes its rise in something which consists of these]; as lovely women, &c.

c. For an agreeable woman gives pleasure to her husband, and, therefore, [is known to be mainly made up of, or] partakes of the quality of 'Goodness;' the indiscreet one gives pain to him, and, therefore, partakes of the quality of 'Foulness;' and she who is separated [and perhaps forgotten,] occasions indifference, and so partakes of the quality of Darkness.'

d. And the appropriate refutation [of any objection], in this case, is [the principle], that it is fitting that the qualities of the effect should be [in every case,] in conformity with the qualities of the cause.

e. Now he states how, in a different way, we have [the evidence of] inference for [the existence of] Soul, which is void of the relation of cause and effect that has been mentioned, [in the four preceding aphorisms, as existing between Nature and its various products]:

Aph. 66.* [The existence] of Soul [is inferred] from the fact that the combination [of the principles of Nature into their various effects] is for the sake of another [than unintelligent Nature, or any of its similarly unintelligent products].

a. 'Combination,' *i.e.*, conjunction, which is the cause [of all products; these resulting from the conjunction of their constituent parts]. Since whatever has this quality, as Nature, Mind, and so on [unlike soul, which is *not* made up of parts], is for the sake of some other; for this reason it is understood that soul exists: such is the remainder, [required to complete the aphorism].

b. But the application [of the argument, in this particular case, is as follows]:

(1) The thing in question, *viz.*, Nature the 'Great one,' with the rest [of the aggregate of the unintelligent], has, as its fruit [or end], the [mundane] experiences and the [eventual] Liberation of some other than itself:

(2) Because it is a combination [or *compages*];

(3) [And every combination,] as a couch, or a seat, or the like, [is for another's use, not for its own; and its several component parts render no mutual service].

c. Now, in order to establish that it is the cause of all [products], he establishes the *eternity* of Nature (*prakriti*):

Aph. 67.* Since the root has no root, the root [of all] is rootless.

a. Since 'the root' (*mula*), *i.e.*, the cause of the twenty-three principles, [which, with soul and the root itself, make up the twenty-five realities recognized in the Samkhya,] 'has no root,' *i.e.*, has no cause, the 'root,' *viz.*, Nature (*pradhana*), is 'rootless,' *i.e.*, void of root. That is to say, there is no other cause of Nature; because there would be a *regressus in infinitum*, [if we were to suppose another cause, which, by parity of reasoning, would require another cause; and so on without end].

b. He states the argument [just mentioned] in regard to this, [as follows]:

Aph. 68.* Even if there be a succession, there is a halt at some *one* point; and so it is merely a name [that we give to the point in question, when we speak of the *root* of things, under the the name of 'Nature'].

a. Since there would be the fault of *regressus in infinitum*, if there were a succession of causes,—another cause of Nature, and another [cause] of that one, again,—there must be, at last, a halt, or conclusion, at some one point, somewhere or other, at some one, uncaused, eternal thing. Therefore, that at which we stop is the *Primal Agency* (*pra-k[iti*); for this [word *prak[iti*, usually and conveniently rendered by the term *Nature*,] is nothing more than a sign to denote the cause which is the *root*: such is the meaning.

b. But then [some Vedanti may object according to this view of matters], the position that there are just twenty-five realities is not made out; for, in addition to the 'Indiscrete' [or primal Nature], which [according to you,] is the cause of Mind, *another* unintelligent principle, named 'Ignorance' [see *Vedanta-sara*, § 21], presents itself. Having pondered this doubt, he declares [as follows]:

Aph. 69.* Alike, in respect of Nature, [and of both Soul and Nature, is the argument for the uncreated existence].

a. In the discussion of the Primal Agent [Nature], the cause which is the root [of all products], the same side is taken by us both, the asserter [of the Samkhya doctrine] and the opponent [Vedanti]. This may be thus stated: As there is mention, in Scripture, of the *production* of Nature, so, too, is there of that of *Ignorance*, in such texts as this, *viz.*: 'This Ignorance, which has five divisions, was produced from the great Spirit.'* Hence it must needs be that a figurative production is intended to be asserted, in respect of *one* of these [and not the *literal* production of both; else we should have no root at all]; and, of the two, it is with *Nature* only that a figurative production, in the shape of a manifestation through conjunction with Soul, &c., is congruous. A production [such as that metaphorical one here spoken of,] the characteristic of which is conjunction *is* mentioned; for there is mention of [such] a figurative origination of Soul and Nature, in a passage of the *Kaurma* [*Purana*], beginning, 'Of action [or the Primal Agency], and knowledge [or Soul],' and so on. And, as there is no mention, in Scripture, of the origin of *Ignorance*, as figurative, *it* is *not* from eternity. And Ignorance, which consists of false knowledge, has been declared, in an aphorism of the *Yoga*, to be [not a separate entity, but] 'an affection of the mind.' Hence there is no increase to the [list of the twenty-five] Realities, [in the shape of a twenty-sixth principle, to be styled ignorance].

b. Or [according to another, and more probable, interpretation of the aphorism,] the meaning is this, that the argument is the same in support of both, *i.e.*, of both Soul and Nature: such is the meaning.

c. But then, there being [as has been shown,] a mode of arriving, by inference, at [a knowledge of the saving truth in regard to] Nature, Soul, &c., whence is it that reflexion, in the shape of discrimination [between Soul and Nature], does not take place in the case of *all* [men]? In regard to this point, he states [as follows]:

Aph. 70.* There is no rule [or necessity, that *all* should arrive at the truth]; because those who are privileged [to engage in the inquiry] are of three descriptions.

a. For those privileged [to enguge in the inquiry] are of three descriptions, through their distinction into those who, in reflecting, are dull, mediocre, and best. Of these, by the dull the [Samkhya] arguments are frustrated [and altogether set aside], by means of the sophisms that have been uttered by the *Bauddhas*, &c. By the mediocre they [are brought into doubt, or, in other words,] are made to appear as if there were equally strong arguments on the other side, by means of arguments which really prove the reverse [of what these people employ them to prove], or by arguments which are not true: [see the section on Fallacies in the *Tarka-sangraha*]. But it is only the *best* of those privileged, that reflect in the manner that has been set forth [in our exposition of the process of reflexion which leads to the discriminating of Soul from Nature]: such is the import. But there is no rule that *all* must needs reflect in the manner so set forth: such is the literal meaning.

b. He now, through two aphorisms, defines 'the Great one' and 'Self-consciousness'; [the reader being presumed to remember that Nature consists of the three 'Qualities' in equipoise, and to be familiar with the other principles, such as the 'Subtile elements' (see § 61)]:

Aph. 71.* The first product [of the Primal Agent, Nature], which is called 'the Great one,' is Mind.

a. 'Mind' (*manas*). 'Mind' [is so called], because its function is 'thinking' (*manana*). By 'thinking' is here meant 'judging' (*nishchaya*). That of which this is the function is 'intellect' (*buddhi*); and *that* is the first product, that called 'the Great one' (*mahat*): such is the meaning.

Aph. 72.* 'Self-consciousness' is that which is subsequent [to Mind.]

a. 'Self-consciousness,' the function of which is a conceit [that '*I* exist,' '*I* do this, that, and the other thing'], is that which is subsequent: that is to say, 'Self-consciousness' is the next after 'the Great one' [§ 71].

b. Since 'Self-consciousness' is that whose function is a conceit [which brings out the *Ego*, in every case of cognition, the matter of which cognition would, else, have lain dormant in the bosom of Nature, the formless Objective], it therefore follows that the others [among the phenomena of mundane existence,] are effects of this [Self-consciousness]; and so he declares [as follows]:

Aph. 73.* To the others it belongs to be products thereof, [*i.e.*, of Self-consciousness].

a. 'To be products thereof,' *i.e.*, to be products of Self-consciousness: that is to say, the fact of being products thereof belongs to the others, the eleven 'Organs' (*indriya*), the five 'Subtile elements,' and, mediately, to the [gross] Elements, also, the products of the Subtile elements.

b. But then, if it be thus [some one may say], you relinquish your dogma, that Nature is the cause of the whole world. Therefore he declares [as follows]:

Aph. 74.* Moreover, mediately, through that [*i.e.*, the 'Great one' (§ 71)] the first [cause, *viz.*, Nature,] is the cause [of all products]; as is the case with the Atoms, [the causes, though not the immediate causes, of jars, &c.].

a. 'Moreover, mediately,' *i.e.*, moreover, not in the character of the immediate cause, 'the first,' *i.e.*, Nature, is the Cause of

'Self-consciousness' and the rest, [mediately,] through 'the Great one' and the rest; as, in the theory of the *Vaisheshikas*, the Atoms are the cause of a jar, or the like, only [mediately,] through combinatians of two atoms, and so on: such is the meaning.

b. But then, since, also, both Nature and Soul are eternal, which of them is [really] the cause of the creation's commencing? In regard to this, he declares [as follows]:

Aph. 75.* While both [Soul and Nature] are antecedent [to all products], since the one [*viz.*, Soul,] is devoid [of this character of being a cause], it is applicable [only] to the other of the two, [*viz.*, Nature].

a. That is to say: 'while both,' *viz.*, Soul and Nature, are preexistent to every product, still, 'since the one,' *viz.*, Soul, from the fact of its not being modified [into anything else, as clay is modified into a jar], must be 'devoid,' or lack the nature of a cause, 'it is applicable,' *i.e.*, the nature of a cause must belong, to the *other* of the two.

b. But then [some one may say], let *Atoms* alone be causes; since there is no dispute [that *these* are causal]. In reply to this, he says:

Aph. 76.* What is limited cannot be the substance of all [things].

a. That which is limited cannot be the substance of all [things]; as yarn cannot be the [material] cause of a jar. Therefore it would [on the theory suggested,] be necessary to mention separate causes of [all] things severally; and it is simpler to assume a single cause. Therefore Nature alone is the cause. Such is the meaning.

b. He alleges Scripture in support of this:

Aph. 77.* And [the proposition that Nature is the cause of all is proved] from the text of Scripture, that the origin [of the world] is therefrom, [*i.e.*, from Nature].

a. An argument, in the first instance, has been set forth [in § 76; for, till argument fails him, no one falls back upon authority]. Scripture, moreover, declares that Nature is the cause of the world, in such terms as, 'From Nature the world arises,' &c.

b. But then [some one may say], a jar which antecedently did not exist is seen to come into existence. Let, then, *antecedent non-existence* be the cause [of each product]; since this is an invariable antecedent, [and, hence, a cause; 'the invariable antecedent being denominated a cause,' if Dr. Brown, in his 6th lecture, is to be trusted]. To this he replies:

Aph. 78.* A thing is not made out of nothing.

a. That is to say: it is not possible that out of nothing, *i.e.*, out of a nonentity, a thing should be made, *i.e.*, an entity should arise. If an entity were to arise out of a nonentity, then, since the character of a cause is visible in its product, the *world*, also, would be unreal: such is the meaning.

b. Let the world, too, *be* unreal: what harm is that to us? [If any ask this,] he, therefore, declares [as follows]:

Aph. 79.* It [the world] is not unreal; because there is no fact contradictory [to its reality], and because it is not the [false] result of depraved causes, [leading to a belief in what ought not to be believed].

a. When there is the notion, in regard to a shell [of a pearl-oyster, which sometimes glitters like silver], that it is silver, its being silver is contradicted by the [subsequent and more correct] cognition, that this is *not* silver. But, in the case in question [that of the world regarded as a reality], no one ever has the cognition, 'This world is *not* in the shape of an entity,' by which [cognition, if any one ever really had such,] it being an entity might be opposed.

b. And it is held that that is false which is the result of a *depraved* cause; *e.g.*, some one's cognition of a [white] conch-shell as *yellow*, through such a fault as the jaundice, [which depraves his eye-sight]. But, in the case in question, [that of the world regarded as a reality], there is not such [temporary or occasional] depravation [of the senses]; because all, at all times, cognize the world as a reality. Therefore the world is *not* an unreality.

c. But then [some one may suggest], *let* a nonentity be the [substantial] cause of the world; still the world will not [necessarily, therefore,] be unreal. In regard to this, he declares [as follows]:

Aph. 80.* If it [the substantial cause,] be an entity, then this would be the case, [that the product would be an entity], from its union [or identity] therewith; [but] if [the cause be] a nonentity, then how could it possibly be the case [that the product would be real], since *it* is a nonentity, [like the cause with which it is united, in the relation of identity]?

a. If an *entity* were the substantial cause [of the world], then, since [it is a maxim that] the qualities of the cause present themselves in the product, 'this would be the case,' *i.e.*, it would be the case that the product was real, 'because of union therewith,' *i.e.*, because of the union [of the product] with the reality [which is its subatratum]. [But,] since, [by parity of reasoning], if a *nonentity* [were the substantial cause], the world would be a nonentity, then, by reason of its being a nonentity, *i.e.*, by reason of the world's being [on that supposition,] necessarily a non-entity, [like its supposed cause], how could this be the case, [that it would be *real*]?

b. But then [a follower of the *MimaEsa* may say], since [it would appear that] nonentity can take no shape but that of nonentity, let *works* alone be the cause of the world. What need have we of the hypothesis of 'Nature'? To this he replies:

Aph. 81.* No; for *works* are not adapted to be the *substantial* cause [of any product].

a. Granting that 'the uneeen' [merit or demerit arising from actions] may be an *instrumental* cause, [in bringing about the mundane condition of the agent], yet we never see merit or demerit in the character of the *substantial* cause [of any product]: and our theories ought to show deference to our experience. 'Nature' is to be accepted; because Liberation arises [see § 56,3 and § 83,] from discerning the distinction between Nature and the Soul.

b. But then [some one may say], since Liberation can be attained by undertaking the things directed by the Veda, what occasion is there for [our troubling ourselves about] *Nature*? To this he replies:

Aph. 82.* The accomplishment thereof [*i.e.*, of Liberation] is not moreover, through Scriptural rites: the chief end of man does not consist in this [which is gained through such means]; because, since this consists of what is accomplished through *acts*, [and is therefore, a *product*, and not *eternal*], there is [still left impending over the ritualist,] the liability to repetition of births.

a. 'Scriptural means,' such as sacrifices, [are so called], because they are heard from [the mouth of the instructor in] Scripture. Not thereby, moreover, is 'the accomplishment thereof,' *i.e.*, the accomplishment of Liberation; 'because one is liable to repetition of births, by reason of the fact that it [the supposed Liberation,] was accomplished by *means*,' *i.e.*, because the [thus far] liberated [soul] is still liable to repetition of births, inasmuch as this [its supposed Liberation,] is not *eternal*, [just] because it is [the result of] *acts*. For *this* reason, the chief end of man does not consist in this, [which is gained through ritual observances].

b. He shows what *does* constitute the chief end of man:

Aph. 83.* There is Scripture for it, that he who has attained to discrimination, in regard to these [*i.e.* Nature and Soul], has no repetition of births.

a. 'In regard to these,' *i.e.*, in regard to Nature and Soul, of him who has attained to discrimination, there is a text declaring, that, in consequence of his knowledge of the distinction, there will be no repetition of births; the text, *viz.*, '*He* does not return again,' &c.

b. He states an objection to the opposite view:

Aph. 84.* From pain [occasioned, *e.g.*, to victims in sacrifice] must come pain [to the sacrificer, and not *liberation* from pain]; as there is not relief from chilliness, by affusion of water.

a. If Liberation were to be effected by *acts*, [such as sacrifices], then, since the acts involve a variety of pains, Liberation itself [on the principle that every effect includes the qualities of its cause,] would include a variety of pains; and it would be a grief, from the fact that it must eventually end: for, to one who is distressed by chilliness the affusion or water does not bring liberation from his chilliness, but, rather, [additional] chilliness.

b. But then [some one may say], the fact that the act is productive of pain is not the *motive* [to the performance of sacrifice]; but the [real] reason is this, that the act is productive of *things desirable*. And, in accordance with this, there is the text, 'By means of acts [of sacrifice] they may partake of immortality,' &c. To this he replies:

Aph. 85.* [Liberation cannot arise from acts]; because whether the end be something desirable, or undesirable, [and we admit that the *motive* of the sacrifice is not the giving pain to the victim], this makes no difference in regard to its being the result of *acts*, [and, therefore, not eternal, but transitory].

a. Grant that pain is not what is [intended] to be accomplished by works done without desire, [on the part of the virtuous sacrificer], still, though there *is* a difference [as you contend,] between [an act done to secure] something enjoyable and an act done without reference to enjoyment, this makes no difference with repect to the fact of the Liberation's being produced by *acts*, [which, I repeat, *permanent* Liberation cannot be]: there must still again be pain; for it [the Liberation supposed to have been attained through works,] must be perishable, because it is a *production*. The text which declares that works done without desire are instruments of Liberation has reference to *knowledge*, [which, I grant, may be gained by such means]; and Liberation comes through knowledge; so that these [works] are instruments of Liberation *mediately*: [but you will recollect that the present inquiry regards the *immediate* cause].

b. [But then, some one may say], supposing that Liberation may take place [as you Sankhyas contend,] through the knowledge of the distinction between Nature and Soul, still, since, from the perishableness [of the Liberation effected by *this* means, as well as any other means], mundane life may return, were both on an equality, [*we*, whose Liberation you Sankhyas look upon as transitory, and you Sankhyas, whose

Liberation we, again, look upon as being, by parity of reasoning, in much the same predicament]. To this he replies:

Aph. 86.* Of him who is essentially liberated, his bonds having absolutely perished, it [*i.e.*, the fruit of his saving knowledge,] is absolute: there is no parity [between his case and that of him who relies on works, and who may thereby secure a temporary sojourn in Paradise, only to return again to earth].

a. Of him 'who is essentially liberated,' who, in his very essence, is free, there is the destruction of bondage. The bond [see § 56,1] is Non-discrimination [between Nature and Soul]. By the removal thereof there is the destruction, the annihilation, of Non-discrimination: and how is it possible that there should again be a return of the mundane state, when the destruction of Non-discrimination is *absolute*? Thus there is no [such] similarity, [between the two cases, as is imagined, by the objector, under § 85. *b*.].

b. It has been asserted [in § 61,] that there is a class of twenty-five [things which are realities], and, since these cannot be ascertained [or made out to be *true*], except by *proof*, therefore he displays this; [*i.e.*, he shows what he *means* by proof]:

Aph. 87.* The determination of something not [previously] lodged in both [the Soul and the Intellect], nor in one or other of them, is 'right notion' (*prama*). What is, in the highest degree, productive thereof [*i.e.*, of any given 'right notion'], is that; [*i.e.*, is what we mean by proof, or evidence, (*pramana*)].

a. 'Not lodged,' *i.e.*, not deposited in 'one rightly cognizing' (*pramat[i*); in short, not previously known. The 'determination,' *i.e.*, the ascertainment [or right comprehension] of such a thing, or reality, is 'right notion'; and, whether this be an affection 'of *both*,' *i.e.*, of Intellect, and also of Soul [as some hold that it is], or of only one or other of the two, [as others hold,] *either* way, 'what is, in the highest degree, productive' of this 'right notion' is [what we term proof, or] evidence, (*pramana*): such is the definition of evidence in general; [the definition of its several species falling to beconsidered hereafter]: such is the meaning.

b. It is with a view to the exclusion of Memory, Error, and Doubt, in their order, that we employ [when speaking of the result of evidence,] the expressions 'not previously known' [which excludes things remembered], and 'reality' [which excludes mistakes and fancies], and 'discrimination,' [which excludes doubt].

c. In regard to this [topic of knowledge and the sources of knowledge], if 'right notion,' is spoken of as located in the *Soul* [see § 87. *a*.], then the [proof, or] evidence is an affection of the *Intellect*. If [on the other hand, the 'right notion' is spoken of as] located in the Intellect, in the shape of an affection [of that the affections of which are mirrored by the Soul], then it [the proof, or evidence, or whatever we may choose to call that from which 'right notion' results,] is just the conjunction of an organ [with its appropriate object; such conjunction giving rise to sense-perception], &c. But, if *both* the Soul's cognition and the affections of the Intellect are spoken of as [cases of] 'right notion,' then *both* of these aforesaid [the affection of the Intellect, in the first case, and the conjunction of an organ with its appropriate object, &c., in the other case,] are [to receive the name of] proof (*pramana*). You are to understand, that, when the organ of vision, &c., are spoken of as 'evidence,' it is only as being *mediately* [the sources of right knowledge].

d. How many [kinds of] proofs [then,] are there? To this he replies:

Aph. 88.* Proof is of three kinds:1 there is no establishment of more; because, if these be established, then all [that is true] can be established [by one or other of these three proofs].

a. 'Proof is of three kinds;' that is to say, 'perception' (*pratyaksha*), 'the recognition of signs' (*anumana*), and 'testimony' (*shabda*), are the [three kinds of] proofs.

b. But then [some one may incline to say], let 'comparison' [which is reckoned, in the Nyaya, a specifically distinct source of knowledge], and the others [such as 'Conjecture,' &c., which are reckoned, in like manner, in the MimaEsa], also be instruments of right knowledge, [as well as these three], in [the matter of] the discriminating of Nature and Soul: he therefore says, 'because, if those [three] be established,' &c. And, since, if there be the three kinds of proof established, 'everything [that is really true] can be established [by means of them], there is no establishment of more;' no addition to the proofs can be fairly made out; because of the cumbrousness [that sins against the philosophical maxim, that we are not to assume more than is necessary to account for the case]: such is the meaning.

c. For the same reason, Manu, also, has laid down only a triad of proofs, where he says [see the Institutes, Ch. xii., v. 105]: 'By that man who seeks a distinct knowledge of his duty, [these] three [sources of right knowledge] must be well understood, *viz.*, Perception, Inference, and Scriptural authority in its various shapes [of legal institute, &c.].' And 'Comparison,' and 'Tradition' (*aitihya*), and the like, are included under Inference and Testimony; and 'Non-perception' (*anupalabdhi*) and the like are included under Perception; [for the non-perception of an absent jar on a particular spot of ground is nothing else than the perception of that spot of ground *without* a jar on it].

d. He [next] states the definitions of the varieties [of proof, having already (§ 87) given the general definition]:

Aph. 89.* Perception (*pratyakska*) is that discernment which, being in conjunction [with the thing perceived], portrays the form thereof.

a. 'Being in conjunction,' [literally,] 'existing in conjunction;' 'portrays the form thereof,' *i.e.*, assumes the form of the thing with which it is in conjunction [as water assumes the form of the vessel into which it is poured]; what 'discernment,' or affection of the Intellect, [does *this*], that [affection of the Intellect (see Yoga Aphorisms, I., § 5 and § 8. *b.*)] is the evidence [called] Perception: such is the meaning.

b. But then, [some one may say,] this [definition of Perception (§ 89)] does not extend [as we conceive it ought, and presume it is intended, to do,] to the perception, by adepts in the *Yoga*, of things past, future, or concealed [by stone walls, or such intervening things as interrupt ordinary perception]; because there is, here, 'no form of the thing, in *conjunction*' [with the mind of him who perceives it, while absent]: having pondered this doubt, he corrects it by [stating, as follows,] the fact, that this [supernatural sort of perception] is not what he intends to define:

Aph. 90.* It is not a fault [in the definition, that it does not apply to the perceptions of adepts in the *Yoga*]; because that of the adepts in the *Yoga* is not an *external* perception.

a. That is to say: it is only *sense*-perception that is to be here defined; and the adepts of the Yoga do not perceive through the *external* [organs of sense]. Therefore there is no fault [in our definition]; *i.e.*, there is no *failure* to include the perceptions of these; [because there is no *intention* to include them].

b. [But, although this reply is as much as the objector has any right to expect,] he states the real justification [of the definition in question]:

Aph. 91.* Or, there is no fault [in the definition], because of the conjunction, with *causal* things, of that [mystical mind] which has attained exaltation.

a. Or, be it so that the perception of the *Yogi*, also, shall be the thing to be defined; still there is no fault [in our definition, § 89]; it does not fail to extend [to this, also]; since the mind of the *Yogi*, in the exaltation gained from the habitude produced by concentration, *does* come into conjunction with things [as existent] in their causes, [whether or not with the things as developed into products perceptible by the external senses].

b. Here the word rendered 'causal' (*lina*) denotes the things, *not* in conjunction [with the senses], alluded to by the objector [in § 89. *b.*]; for *we*, who assert that effects *exist* [from eternity, in their causes, before taking the shape of effects, and, likewise, in these same causes, when again resolved into their causes], hold that even what is past, &c., still essentially exists, and that, hence, its conjunction [with the mind of the mystic, or the clairvoyant,] is possible.

c. But then, [some one may say,] still this [definition] does not extend to the *Lord's* perceptions; because, since these are from everlasting, they cannot *result* from [emergent] conjunction. To this he replies:

Aph. 92.* [This objection to the definition of Perception has no force]; because it is not proved that there *is* a Lord (*ishwara*).

a. That there is no fault [in the definition of Perception], because there is no proof that there *is* a Lord, is supplied [from § 90].

b. And this demurring to there being any 'Lord' is merely in accordance with the arrogant dictum of [certain] partisans [who hold an opinion not recognized by the majority]. Therefore, it is to be understood, the expression employed is, 'because it is *not proved* that there is a Lord,' but not the expression, 'because there *is no* Lord,'

c. But, on the implication that there *is* a 'Lord,' what we mean to speak of [in our definition of Perception, (§ 89),] is merely the being of the [same] kind with what is produced by conjunction [of a sense-organ with its object; and the perceptions of the 'Lord' may be of the same *kind* with such perceptions, though they were not to come from the same *source*].

d. Having pondered the doubt, '*How* should the Lord not be proved [to exist] by the Scripture and the Law, [which declare his existence]?' he states a dilemma which excludes [this]:

Aph. 93.* [And, further,] it is not proved that he [the 'Lord,'] exists; because [whoever exists must be either free or bound; and], of free and bound, he can be neither the one nor the other.

a. The 'Lord' whom you imagine, tell us, is he free from troubles, &c.? Or is he in bondage through these? Since he is not, cannot be, either the one or the other, it is not proved that there is a 'Lord:' such is the meaning.

b. He explains this very point:

Aph. 94.* [Because,] either way, he would be inefficient.

a. Since, if he were free, he would have no desires, &c., which [as compulsory motives,] would instigate him to create; and, if he were bound, he would be under delusion; he must be [on either alternative,] unequal to the creation, &c. [of this world].

b. But then, [it may be asked,] if such be the case, what becomes of the Scripture-texts which declare the 'Lord?' To this he replies:

Aph. 95.* [The Scriptural texts which make mention of the 'Lord' are either glorifications of the liberated Soul, or homages to the recognized [deities of the Hindu pantheon].

a. That is to say: accordingly as the case may be, *some* text [among those in which the term 'Lord' occurs,] is intended, in the shape of a glorification [of Soul], as the 'Lord,' [as Soul is held to be], merely in virtue of junction [with Nature], to incite [to still deeper contemplation], to exhibit, as what is to be known, the liberated Soul, *i.e.*, absolute Soul in general; and some other text, declaratory, for example, of creatorship, &c., preceded by resolution [to create, is intended] to extol [and to purify the mind of the contemplator, by enabling him to take a part in extolling] the eternity, &c., of the familiarly known Brahma, Vishnu, shiva, or other *non*-eternal 'Lord;' since these, though possessed of the conceit [of individuality], &c., [and, in so far, liable to perish], have immortality, &c., in a secondary sense; [seeing that the Soul, in *every* combination, is immortal, though the combination itself is not so].

b. But then, [some one may say], even if it were thus [as alleged under § 95], what is heard in Scripture, [*viz.*], the fact that it [*viz.*, Soul] is the *governor* of Nature, &c., would not be the case; for, in the world, we speak of government in reference only to modifications [preceded and determined] by resolutions [that so and so shall take place], &c. To this he replies:

Aph. 96.* The governorship [thereof, *i.e.*, of Soul over Nature] is from [its] proximity thereto, [not from its resolving to act thereon]; as is the case with the gem, [the lodestone, in regard to iron].

a. If it were alleged that [its, Soul's] creativeness, or [its] governorship, was through a *resolve* [to create, or to govern], then this objection [brought forward under § 95. *b*.] would apply. But [it is not so; for,] by *us* [Sankhyas,] it is held that the Soul's governorship, in the shape of creatorship, or the like, is merely from [its] *proximity* [to Nature]; 'as is the case with the [lodestone] gem.'

b. As the gem, the lodestone, is attracted by iron merely by proximity, without resolving [either to act or to be acted on], &c., so, by the mere conjunction of the primal Soul, Nature is changed into the principle [called] the 'Great one,' [or Mind, (see § 61. *c*.)]. And in this alone consists [what we speak of as] its acting as *creator* towards that which is superadded to it: such is the meaning

c. And thus it is declared, [in some one of the Puranas]: 'As the iron acts, whilst the gem [the lodestone,] stands void of volition, just so this world is created by a deity who is mere Existence. Thus it is, that there are, in the Soul, both agency [seemingly,] and non-agency, [really]. It is *not* an agent, inasmuch as it is void of volition; [and it *is*] an agent, merely through approximation [to Nature].'

d. In respect of worldly products, also, animal souls overrule, merely through their approximation [to Nature]: so he declares [as follows]:

Aph. 97.* In the case of individual products, also, [the apparent agency] of animal souls [is solely through proximity].

a. 'The agency is solely through proximity:' so much is supplied [from § 96].

b. The meaning is this, that, in the case, also, of particular productions,—the creation, &c., of things individual [as contradistinguished from that of all things in the lump, (see *Vedanta-sara*, § 67)],—animal souls, *i.e.*, souls in which the intellects [of individuals] reflect themselves [see § 99. *a*.], overrule, merely through proximity, but not through any effort; seeing that these [animal souls] are none other than the motionless Thought.

c. But then, [some one may say], if there were no eternal and omniscient 'Lord,' through the doubt of a blind tradition, [in the absence of an intelligently effective guardianship], the *Vedas* would cease to be an authority; [a possibility which, of course, cannot be entertained for an instant]. To this he replies:

Aph. 98.* The declaration of the texts or sense [of the Veda, by Brahma, for example], since *he* knows the truth, [*is* authorative evidence].

a. To complete [the aphorism, we must say], 'since *Hiranyagarbha* [*i.e.*, *Brahma*,] and others [*viz.*, *Vishnu* and *shiva*], are knowers of what is certain, *i.e.*, of what is true, the declaration of the texts or sense of the Vedas, where *these* are the speakers, *is* evidence [altogether indisputable].'

b. But then, if Soul, by its simple proximity [to Nature (§ 96)], is an overruler in a *secondary* sense [only of the term,— as the magnet may be said, is a secondary sense, to draw the iron, while the conviction is entertained, that, actually and literally, the iron draws the magnet],—who is the *primary* [or actual,] overruler? In reference to this, he says:

Aph. 99.* The internal organ, through its being enlightened thereby [*i.e.*, by Soul], is the overruler; as is the iron, [in respect of the magnet].

a. The internal organ, *i.e.*, the understanding, is the overruler, through its fancying itself to be Soul, [as it does fancy,] by reason of its being enlightened by the Soul, through its happening to reflect itself in [and contemplate itself in,] Soul; 'just as the iron,' that is to say, as the attracting iron, though inactive, draws [the magnet], in consequence of [its] mere proximity, [and so acquires magnetism by magnetic induction].

b. He [now, having discussed the evidence that consists in direct perception,] states the definition of inference (*anumana*):

Aph. 100.* The knowledge of the connected [*e.g.*, fire], through perception of the connexion [*e.g.*, of fire with smoke], is inference.

a. That is to say: inference [or conviction of a general truth,] is [a kind of] evidence consisting in a [mental] modification, [which is none other than] the knowledge of the connected, *i.e.*, of the constant accompanier, through the knowledge of the constant accompaniment: by 'connexion' (*pratibandha*) here being meant 'constant attendedness' (*vyapti*); and through the perception thereof [it being that the mind has possession of any general principle].

b. But a conclusion (*anumiti*) is knowledge of the soul; [whilst an Inference, so far forth as it is an instrument in the establishment of knowledge deducible from it, is an affection of the internal organ, or understanding (see § 87. *c.*)]

c. He [next] defines testimony (*shabda*):

Aph. 101.* Testimony [such as is entitled to the name of evidence, is a declaration by one worthy [to be believed].

a. Here 'fitness' means 'suitableness;' and so the evidence which is called 'Testimony' is the knowledge arising from a suitable declaration: such is the meaning. And [while this belongs to the understanding, or internal organ (see § 100. *b.*)] the result is that [knowledge] in the Soul, [which is called,] 'knowledge by hearing' (*shabda-bodha*).

b. He [next] volunteers to tell us what is the use of his setting forth [the various divisions of] evidence:

Aph. 102.* Since the establishment of [the existence of] both [soul and non-soul] is by means of evidence, the decaration thereof [*i.e.*, of the kinds of evidence, has been here made].

a. It is only by means of evidence that both Soul and non-soul are established as being distinct, [the one from the other]: therefore has this, *viz.*, evidence, been here declared: such is the meaning.

b. Among these [several kinds of proof], he [now] describes that one by which, especially, *viz.*, by a proof which is one kind of inference, Nature and Soul are here to be established discriminatively:

Aph. 103.* The establishment of both [Nature and Soul] is by analogy.

a. [Analogy (*samanyato d[ishma*) is that kind of evidence which is employed in the case] where, by the force [as an argument,] which the residence of any property in the subject derives from a knowledge of its being constantly accompanied [by something which it may therefore betoken], when we have had recourse to [as the means of determining this constant accompaniment,] what is, for instance, generically of a perceptible kind, [where, under such circumstances, we repeat,] anything of a *different* kind, *i.e.*, *not* cognizable by the senses, is established; as when, for example, having apprehended a constant accompaniment, [*e.g.*, that an act implies an instrument], by taking into consideration such instruments as axes, &c., which are of earthy and other kinds, a quite heterogeneous, imperceptible, instrument of *knowledge*, *viz.*, [the instrument named] Sense, is established [or inferred to exist]; such is what we mean by Analogy; and it is by *this* [species of inference], that both, [*viz.*,] Nature and Soul are proved [to exist]: such is the meaning.

b. Of these [*viz.*, Nature and Soul,] the argument from analogy for [the existence of] Nature is as follows: the Great Principle [*viz.*, Understanding (see § 61. *c*.)] is formed out of the things [called] Pleasure, Pain, and Delusion, [to the aggregate of which three in equipoise (see § 61) the name of Nature is given]; because, whilst it is [undeniably,] a production, it has the characters of Pleasure, Pain, and Delusion; just as a bracelet, or the like, formed of gold, or the like; [has the characteristic properties of the gold, or the like, and is thereby known to have been formed out of gold, or the like].

c. But, [as regards tht argument from analogy, in proof of the existence] of Soul, [it is, as stated before, under § 66, to the following effect]: Nature is for the sake of *another*; because it is something that acts as a combination; as a house, for instance, [which is a combination of various parts combined for the benefit of the tenant]. In this instance, having gathered, in regard to houses, &c., the fact established on sense-perception, that they exist for the sake of [organized] bodies, for example, something of a different kind therefrom, [*i.e.*, from Nature, *viz.*], Soul, is inferred [by analogy,] as something other than Nature, &c., [which, as being a compound thing, is not designed for itself]: such is the meaning.

d. But then [some one may say], since Nature is eternal, and exertion is habitual to her, [and the result of her action is the bondage of the Soul], there should constantly be experience [whether of pleasure or of pain], and, hence, no such thing as thorough emancipation. To this he replies:

Aph. 104.* Experience [whether of pain or pleasure, ends with the discernment of] Thought, [or Soul, as contradistinguished from Nature].

a. By 'Thought' [we mean] soul. Experience [whether of pain or pleasure,] ceases, on the discerning thereof. As 'antecedent non-existence,' though devoid of a beginning, [see *Tarka-sangraha*, § 92], surceases [when the thing antecedently non-existent begins to be], so, eternal Nature [eternal, as regards the absence of any beginning,] continues [no further than] till the discernment of the difference [between Nature and Soul]; so that experience [whether of pain or pleasure,] does *not* at all times occur: such is the state of the case.

b. [But some one say], if Nature be agent, and Soul experiencer, then it must follow [which seems unreasonable,] that another is the experiencer of [the results of] the acts done by one different. To this he replies:

Aph. 105.* The experience of the fruit *may* belong even to another than the agent; as in the case of food, &c.

a. As it belongs to the cook to prepare the food, &c., and to one who was not the agent, *viz.*, the master, to enjoy the fruit [thereof, *i.e.*, the fruit of the cook's actions], so is the case here, also.

b. Having stated an exoteric principle [which may serve, in practice, to silence, by the *argumentum ad hominem*, him on whose principles it may be valid], he [next] declares his own doctrine, [in regard to the doubt started under § 104. *b.*]:

Aph. 106.* Or, [to give a better account of the matter than that given in § 105], since it is from non-discrimination that it is derived, the notion that the *agent* [soul being mistaken for an agent,] has the fruit [of the act is a wrong notion].

a. The soul is neither an agent nor a patient; but, from the fact that the Great Principle [the actual agent (see § 97. *b.*)] is reflected in it, there arises the *conceit* of its being an agent. 'Or, since it is from non-discrimination;' that is to say, because it is from the failure to discriminate between Nature and Soul, that this takes place, *i.e.*, that conceit takes place, that it is the *agent* that experiences the fruit; [whereas the actual agent is Nature, which, being unintelligent, can experience neither pain nor pleasure].

b. The opposite of this [wrong view, referred to in § 106,] he states [as follows]:

Aph. 107.* And, when the truth is told, there is [seen to be] neither [agency, in Soul, nor experience].

a. 'When the truth is told' [and discerned], *i.e.*, when, by means of evidence, Nature and Soul are *perceived* [in their entire distinctness, one from the other], 'there is neither,' *i.e.*, neither the condition [as regards soul,] of an agent nor that of a patient.

b. Having discussed [the topic of] evidence, he [now] states the distribution of the subject-matter of evidence:

Aph. 108.* [A thing may be] an object [perceptible], and also [at another time,] not an object, through there being, in consequence of great distance, &c., a want of [conjunction of the sense with the thing], or [on the other hand,] an appliance of the sense [to the thing].

a. An object [is a perceived object], through the proximity, or conjunction, of the sense [with the object]. [A thing may be] not an object [perceived], through the want of the sense, *i.e.*, through the want of conjunction [between the sense and what would otherwise be its object]. And [this] want of conjunction [may result] from the junction's being prevented by great distance, &c.

b. [To explain the '&c.,' and to exemplify the causes that may prevent the conjunction, required in order to perception, between the thing and the sense, we may remark, that] it is in consequence of great distance, that a bird [flying very high up] in the sky is not perceived; [then again,] in consequence of extreme proximity, the collyrium located in the eye [is not perceived by the eye itself]; a thing placed in [the inside of, or on the opposite side of,] a wall [is not perceived], in consequence of the obstruction; from distraction of mind, the unhappy, or other [agitated person], does not perceive the thing that is at his side [or under his very nose]; through its subtilty, an atom [is not perceived]; nor is a very small sound, when overpowered by the sound of a drum; and so on.

c. How [or, for which of the possible reasons just enumerated,] comes the imperceptibleness of *Nature*? In regard to this, he declares:

Aph. 109.* Her imperceptibleness arises from [her] subtility.

a. 'Her,' *i.e.*, *Nature's*, imperceptibleness is from subtilty. By subtilty is meant the fact of being difficult to investigate; not [as a Naiyayika might, perhaps, here prefer understanding the term,] the consisting of atoms; for Nature is [not atomic, in the opinion of the Sankhyas, but] all-pervasive.

b. How, then, [it may be asked,] is [the existence of] Nature determined? To this he replies:

Aph. 110.* [Nature exists;] because her existence is gathered from the beholding of productions.

a. As the knowledge of [there being such things as] atoms comes from the beholding of jars, &c., [which are agglomerations], so the knowledge of Nature comes from the beholding of products which have the three Qualities; [(see § 62. *a*.) and the existence of which implies a cause, to which the name of Nature is given, in which these constituents exist from eternity].

b. Some [the Vedantis,] say that the world has *Brahma* as its cause; others [the Naiyayikas], that it has atoms as its cause; but our seniors [the transmitters of the Samkhya doctrine], that it has *Nature* as its cause. So he sets forth a doubt [which might naturally found itself] thereon:

Aph. 111.* If [you throw out the doubt that] it [*viz*., the existence of Nature,] is not established, because of the contradiction of asserters [of other views, then you will find an answer in the next aphorism].

a. 'Because of the the contradiction of asserters [of the Vedanta or Nyaya], *it* is not established,' *i.e.*, Nature [as asserted by the Sankhyas,] is not established.

b. But then, [to set forth the objection of these counter-asserters], if a product existed antecedently to its production [as that product], *then* an eternal Nature [such as you Sankhyas contend for,] would be proved to exist as the [necessary] substratum thereof; since you will declare that a cause is inferred only as the [invariable] accompanier of an effect; but it is denied, by us asserters [of the Vedanta, &c.], that the effect *does* exist [antecedently to its production; well,] *if* [this doubt be thrown out]: such is the meaning [of the aphorism].

c. He states [his] doctrine [on this point]:

Aph. 112.* Still, since each [doctrine] is established in the opinion of each, a [mere unsupported] denial is not [decisive].

a. If one side were disproved merely by the dissent of the opponent, then [look you,] there is dissent against the other side, too: so how could *it* be established? If the one side is established by there being inevitably attendant the recognition of the constant accompanier, on the recognition of that which is constantly accompanied [by it], it is the same with *my* [side], also: therefore [my] inference from effect [to cause] is not to be denied [in this peremptory fashion].

b. Well, then, [the opponent may say], let [the inference of] cause from effect be granted; how is it that this [cause] is *Nature*, and nothing else, [such as Atoms, for instance]? To this he replies:

Aph. 113.* Because [if we were to infer any other cause than Nature,] we should have a contradiction to the threefold [aspect which things really exhibit].

a. Quality is threefold [see § 61. *a*.], *viz*., Goodness, Passion, and Darkness: there would be a contradiction to *these*: such is the meaning.

b. The drift here is as follows: If the character of cause [of all things around us] belonged to Atoms, or the like, then there would be a contradiction to the fact of being an aggregate of pleasure, pain, and delusion, which is recognizable in the world; [because nothing, we hold, can exist in the effect, which did not exist in the cause, and pleasure, pain, &c., are no properties of Atoms].

c. He now repels the doubt as to whether the production of an effect is that of what existed [antecedently], or of what did not exist:

Aph. 114.* The production of what is no entity, as a man's horn, does not take place.

a. Of that which, like the horn of a man, is not an entity, even the production is impossible: such is the meaning. And so the import is, that that effect alone which [antecedently] exists is [at any time] produced.

b. He states an argument why an effect must be some [previously existent] entity:

Aph. 115.* Because of the rule, that there must be some material [of which the product may consist].

a. And only when both are extant is there, from the presence of the cause, the presence of the effect. Otherwise, everywhere and always, every [effect] might be produced; [the presence of the cause being, on the supposition, superfluous]. This he insists upon [as follows]:

Aph. 116.* Because everything is not possible everywhere and always [which might be the case] if materials could be dispensed with].

a. That is to say: because, in the world, we see that everything is *not* possible, *i.e.*, that everything is *not* produced; 'everywhere,' *i.e.*, in every place; 'always,' *i.e.*, at all times.

b. For the following reason, also, he declares, there is no production of what existed not [antecedently]:

Aph. 117.* Because it is that which is competent [to the making of anything] that makes what is possible, [as a product of it].

a. Because the being the material [of any future product] is nothing else than the fact of [being it, *potentially*, *i.e.*, of] having the competency to be the product; and [this] competency is nothing else than the product's condition as that of what has not yet come to pass: therefore, since 'that which is competent,' *viz*., the cause, makes the product which is 'possible' [to be made out of it], it is not of any *nonentity* that the production takes place, [but of an entity, whose *esse*, antecedently, was *possibility*]: such is the meaning.

b. He states another argument:

Aph. 118.* And because it [the product,] is [nothing else than] the cause, [in the shape of the product].

a. It is declared, in Scripture, that, previously to production, moreover, there is no difference between the cause and its effect; and, since it is thereby settled that a product is an entity, production is not of what [previously] existed not: such is the meaning.

b. He ponders a doubt:.

Aph. 119.* If [it be alleged that] there is no possibility of that's *becoming* which already *is* [then the answer will be found in the next aphorism].

a. That is to say: but then, if it be thus [that every effect exists antecedently to its production], since the effect [*every* effect,] must be eternal [without beginning], there is no possibility of [or room for] the adjunction of *becoming*, the adjunction of *arising*, in the case of a product which is [already, by hypothesis,] in the shape of an entity; because the employment of [the term] '*arising*' [or the fact of being produced] has reference solely to what did *not* exist [previously]; if this be urged: such is the meaning.

b. He declares the doctrine [in regard to this point]:

Aph. 120.* No; [do not argue that what *is* cannot become; for] the employment and the non-employment [of the term 'production'] are occasioned by the *manifestation* [and the non-manifestation of what is spoken of as produced, or not].

a. 'No;' the view stated [in § 119] is not the right one: such is the meaning.

b. As the whiteness of white cloth [which has become] dirty is brought manifestly out by means of washing, &c., so, by the operation of the potter, is the pot brought into manifestness; [whereas], on the blow of a mallet, it becomes hidden, [and no longer appears as a *pot*].

c. And manifestation [is no fiction of ours; for it] is seen; for example, that of oil, from sesamum-seeds, by pressure; of milk, from the cow, by milking; of the statue, which resided in the midst of the stone, by the operation of the sculptor; of husked rice, from rice in the husk, by threshing; &c.

d. Therefore, the employment and the non-employment of the [term] 'the *production* of an effect' are dependent on *manifestation*, dependent on the manifestation of the *effect*: that is to say, the employment of [the term] 'production' is in consequence of the manifestation [of what is spoken of as produced]; and the non-employment of [the term] 'production' is in consequence of there being no manifestation [of that which is, therefore, not spoken of as produced]; but [the employment of the term 'production' is] not in consequence of that's becoming an entity which was not an entity.

e. But if [the employment of the term] 'production' is occasioned by [the fact of] *manifestation*, by what is occasioned [the employment of the term] *destruction*? To this he replies:

Aph. 121.* Destruction [of anything] is the resolution [of the thing spoken of as destroyed,] into the cause [from which it was produced].

a. The resolution, by the blow of a mallet, of a jar into its cause [*i.e.*, into the particles of clay which constituted the jar], to *this* are due both [the employment of] the term 'destruction,' and the kind of action [or behaviour] belonging to anything [which is termed its destruction].

b. [But some one may say], if there were [only] a resolution [of a product into that from which it arose], a resurrection [or À±»133µ½µÃ¯±] of it might be seen; and this is *not* seen: well [we reply], it is not seen by blockheads; but it *is* seen by those who can discriminate. For example, when thread is destroyed, it is changed into the shape of earth [as when burned to ashes]; and the earth is changed into the shape of a cotton-tree; and this [successively] changes into the shape of flower, fruit, and thread [spun again from the fruit of the cotton-plant]. So is it with all entities.

c. Pray [some one may ask], is [this] *manifestation* [that you speak of under § 120] something real, or something not real? If it be something real [and which, therefore, never anywhere ceases to be], then [all] effects [during this constant manifestation] ought constantly to be *perceived*; and, if it be *not* real, then there would be the absence of [all] products, [in the absence of all manifestation. Manifestation, therefore, must be something *real*; and] there must be [in order to give rise to it,] another manifestation of it, and of this another; [seeing that a *manifestation* can be the result of nothing else than a manifestation, on the principle that an effect consists of neither more nor less than its cause]; and thus we have a *regressus in infinitum*. To this he replies:

Aph. 122.* Because they seek each other reciprocally, as is the case with seed and plant, [manifestation may generate manifestation, from eternity to eternity].

a. Be it so, that there are thousands of manifestations; still there is no fault; for there *is* no starting-point; as is the case with seed and plant, [which people may suppose to have served, from eternity, as sources, one to another, reciprocally].

b. He states another argument:

Aph. 123.* Or, [at all events, our theory of 'manifestation' is as] blameless as [your theory of] 'production.'

a. Pray [let us ask], is *production* produced, or is it not? If it is produced, then of this [production of production] there must be production; so that there is a *regressus in infinitum*, [such as you allege against *our* theory, under § 121. *c.*] If it be *not* produced, then, pray, is this because it is *unreal*, or because it is eternal? If because it is unreal, then production never is at all; so that it would never be perceived, [as you allege that it is]. Again, if [production is not something produced,] because it is *eternal*, then there would be at, all times, the production of [all possible] effects, [which you will scarcely pretend is the case]. Again, if you say, since 'production' itself *consists* of production, what need of supposing an ulterior production [of production]? then, in like manner, [*I* ask,] since 'manifestation' itself *consists* of manifestation, what need of supposing an ulterior manifestation [of manifestation]? The view which you hold on this point is *ours*, also; [and thus every objection stated or hinted under § 121. *c*., is capable of being retorted].

b. He [now] states the community of properties [that exists] among the products of Nature, mutually:

Aph. 124.* [A product of Nature is] caused, uneternal, not all-pervading, mutable, multitudinous, dependent, mergent.

a. 'Caused,' *i.e.* having a cause. 'Uneternal,' *i.e.*, destructible. 'Not all-pervading,' *i.e.*, not present everywhere. 'Mutable,' *i.e.*, distinguished by the acts of leaving [one form], and assuming [another form], &c. It [the soul,] leaves the body it has assumed, [and, probably, takes another]; and bodies, &c., move [and are mutable, as is notorious]. 'Multitudinous,' *i.e.*, in consequence of the distinction of souls; [every man, *e.g.*, having a separate body]. 'Dependent,' [*i.e.*,] on its cause. 'Mergent,' that is to say, it [*i.e.*, every product, in due time,] is resolved into that from which it originated.

b. [But, some one may say], if realities be the twenty-five [which the Sankhyas enumerate (see § 61), and no more], pray, are such common operations as knowing, enjoying, &c., absolutely *nothing*; you accordingly giving up what you *see*, [in order to save an hypothesis with which what you see is irreconcilable]? To this he replies:

Aph. 125.* There is the establishment of these [twenty-four 'Qualities' of the *Nyaya*, which you fancy that we do not recognize, because we do not explicitly enumerate them], either by reason that these ordinary qualities [as contradistinguished from the *three* Qualities of the Samkhya], &c., are, in reality, nothing different; or [to put it in another point of view,] because they are hinted by [the term] Nature, [in which, like our own three Qualities, they are implied].

a. Either from their being nothing different from the twenty-four principles, 'in reality,' truly, quite evidently,—since the character of these [twenty-four] fits the ordinary qualities, &c., [which you fancy are neglected in our enumeration of things,]—' there is the establishment of these,' *i.e.*, there is their establishment [as realities,] through their being implied just in those [twenty-four principles which are explicitly specified in the Samkhya].

b. The word 'or' shows that there is another alternative [reply, in the aphorism, to the objection in question]. 'Or because they are hinted by [the term] Nature;' that is to say, the qualities, &c. [such as Knowledge], are established [as realities], just because they are hinted by [the term] Nature, by reason that [these] qualities are, mediately, products of Nature; for there is no difference between product and cause. But the omission to mention them [explicitly] is not by reason of their not being at all.

c. He [next] mentions the points in which Nature and [her] products agree:

Aph. 126.* Of both [Nature and her products] the fact that they consist of the three Qualities [§ 61. *a*.], and that they are irrational, &c., [is the common property].

a. Consisting of the three qualities, and being irrational, [such in the meaning of the compound term with which the aphorism commences]. By the expression '&c.' is meant [their] being intended for *another*, [see § 66]. 'Of both,' *i.e.*, of the cause [*viz*., Nature], and of the effects [*viz*., all natural products]. Such is the meaning:

b. He [next] states the mutual differences of character among the three Qualities which [see § 61] are the [constituent] parts of Nature:

Aph. 127.* The Qualities [§ 62] differ in character mutually by pleasantness, unpleasantness, lassitude, &c., [in which forms, severally, the Qualities present themselves].

a. 'Pleasantness,' *i.e.*, Pleasure. By the expression '&c.' is meant Goodness (*sattwa*), which is light [*i.e.*, not heavy,] and illuminating. 'Unpleasantness,' *i.e.*, Pain. By the expression '&c.' [in reference to this,] is meant Passion (*rajas*), which is urgent and restless. 'Lassitude,' *i.e.*, stupefaction. By the expression '&c.' is meant Darkness (*tamas*), which is heavy and enveloping. It is by these habits that the Qualities, *viz*., Goodness, Passion, and Darkness, differ: such is the remainder, [required to complete the aphorism].

b. At the time of telling their differences, he tells in what respects they agree:

Aph. 128.* Through Lightness and other habits the Qualities mutually agree and differ.

a. The meaning is as follows: the enunciation [in the shape of the term *laghu*, 'light,' is not one intended to call attention to the concrete, *viz.*, what things are light, but] is one where the abstract [the nature of light things, *viz.*, 'lightness' (*laghutwa*)] is the prominent thing. 'Through Lightness and other habits,' *i.e.*, through the characters of Lightness, Restlessness, and Heaviness, the Qualities differ. Their *agreement* is through what is hinted by the expression 'and other.' And this consists in their mutually predominating [one over another, from time to time], producing one another, consorting together, and being reciprocally present, [one in another], for the sake of Soul.

b. By [the expressions, in § 124,] 'caused,' &c., it is declared that the 'Great one' [or Mind], &c., are *products*. He states the proof of this:

Aph. 129.* Since they are other than both [Soul and Nature, the only two uncaused entities], Mind and the rest are products; as is the case with a jar, or the like.

a. That is to say: like a jar, or the like, Mind and the rest are products; because they are something other than the two which [alone] are eternal, *viz.*, Nature and Soul.

b. He states another reason:

Aph. 130.* Because of [their] measure, [which is a limited one].

a. That is to say: [Mind and the rest are products]; because they are limited in measure; [whereas the only two that are uncaused, *viz.*, Nature and Soul, are unlimited].

b. He states another argument:

Aph. 131.* Because they conform [to Nature].

a. [Mind and the rest are products]; because they well [follow and] correspond with Nature; *i.e.*, because the Qualities of Nature [§ 61] are seen in all things: [and it is a maxim, that what is in the effect was derived from the cause and implies the cause].

b. He states the same thing, [in the next aphorism]:

Aph. 132.* And, finally, because it is through the power [of the cause alone, that the product can do aught].

a. It is by the power of its cause, that a product energizes, [as a chain restrains an elephant, only by the force of the iron which it is made of]; so that Mind and the rest, being [except through the strength of Nature,] powerless, produce *their* products in subservience to Nature. Otherwise, since it is their habit to energize, they would at all times produce their products, [which it will not be alleged that they do].

b. And the word *iti*, in this place, is intended to notify the completion of the set of [positive] reasons [why Mind and the others should be regarded as products].

c. He [next] states [in support of the same assertion,] the argument from negatives, [*i.e.*, the argument drawn from the consideration as to what becomes of Mind and the others, when they are *not* products]:

Aph. 133.* On the quitting thereof [quitting the condition of product], there is Nature, or Soul, [into one or other of which the product must needs have resolved itself].

a. Product and non-product; such is the pair or alternatives. 'On the quitting thereof;' *i.e.*, when Mind and the rest quit the condition of product, Mind and the rest [of necessity] enter into Nature, or Soul; [these two alone being non-products].

b. [But perhaps some one may say, that] Mind and the rest may exist quite independently of the pair of alternatives [just mentioned]. In regard to this, he declares [as follows]:

Aph. 134.* If they were other than these two they would be void; [seeing that there is nothing self-existent, besides soul and Nature].

a. If Mind and the rest were 'other than these two,' *i.e.*, than product or non-product [§ 133], they would be in the shape of what is 'void,' *i.e.*, in the shape of nonentity.

b. Well now, [some one may say,] why should it be under the character of a *product*, that Mind and the rest are a sign of [there being such a principle as] Nature? They may be [more properly said to be] a sign, merely in virtue of their not *occurring apart* from it. To this he replies:

Aph. 135.* The cause is inferred from the effect, [in the case of Nature and her products]; because it accompanies it.

a. That [other relation, other than that of material and product, which you would make out to exist between Nature and Mind,] exists, indeed, where the nature [or essence] of the cause is not seen in the effect; as [is the case with] the inference, from the rising of the moon, that the sea is swollen [into full tide; rising, with maternal affection, towards her son who was produced from her bosom on the occasion of the celebrated Churning of the Ocean. Though the swelling of the tide does not occur apart from the rising of the moon, yet here the cause, moon-rise, is not seen in the effect, tide; and, consequently, though we infer the effect from the cause, the cause could not have been inferred from the effect]. But, in the present case, since we see, in Mind and the rest, the characters of Nature, the cause *is* inferred from the effect. 'Because it accompanies it,' *i.e.*, because, in Mind and the rest, we see the properties of Nature, [*i.e.*, Nature herself actually present; as we see the clay which is the cause of a jar, actually present in the jar].

b. [But it may still be objected,] if it be thus, then let that principle itself, the 'Great one' [or Mind], be the cause of the world: what need of *Nature*? To this he replies:

Aph. 136.* The indiscrete, [Nature, must be inferred] from its [discrete and resolvable] effect. [Mind], in which are the three Qualities, [which constitute Nature].

a. 'It is resolved;' such is the import of [the term] *linga*, [here rendered] 'effect.' From that [resolvable effect], *viz.*, the 'Great principle' [or Mind], in which are the three Qualities, Nature must be inferred. And that the 'Great principle,' in the shape of ascertainment [or distinct intellection], is discrete [or limited] and perishable, is established by direct observation. Therefore [*i.e.*, since Mind, being perishable, must be resolvable into something else,] we infer that into which it is resolvable, [in other words, its 'cause,' here analogously termed *lingin*, since 'effect' has been termed *linga*].

b. But then, [some one may say], still something quite different may be the cause [of all things]: what need of [this] *Nature* [of yours]? In regard to this, he remarks [as follows]:

Aph. 137.* There is no denying that it [Nature,] *is*; because of its effects, [which will be in vain attributed to any other source].

a. Is the cause of this [world] a product, or not a product? If it were a product, then, the same being [with equal propriety to be assumed to be] the case with *its* cause, there would be a *regressus in infinitum*. If effects be from any *root* [to which there is nothing antecedent], then *this* is that [to which we give the name of *Nature*]. 'Because of its effects,' that is to say, because of the effects of Nature. There is no denying 'that it is,' *i.e.*, that *Nature* is.

b. Be it so, [let us grant,] that Nature *is*; yet [the opponent may contend,] *Soul* positively cannot be; for [if the existence of causes is to be inferred from their products, Soul cannot be thus demonstrated to exist, seeing that] it has *no* products. In regard to this, he remarks [as follows]:

Aph. 138.* [The relation of cause and effect is] not [alleged as] the means of establishing [the existence of Soul]; because, as is the case with [the disputed term] 'merit,' there is no dispute about there being such a kind of thing; [though *what* kind of thing *is* matter of dispute].

a. There is no dispute about 'there being such a kind of thing,' *i.e.*, as to there being Soul, simply; [since everybody who does not talk stark nonsense must admit a Soul, or *self*, of *some* kind]; for the dispute is [not as to its *being*, but] as to its peculiarity [of being], as [whether it be] multitudinous, or sole, all-pervading, or *not* all-pervading, and so forth; just as, in every [philosophical system, or] theory, there is no dispute as to [there being something to which may be applied the term] 'merit' (*dharma*); for the difference of opinion has regard to the particular kind of [thing,—such as sacrifices, according to the

MimaEsa creed, or good works, according to the Nyaya,—which shall be held to involve] 'merit.'

b. 'Not the means of establishing' that [*viz.*, the existence of soul]; *i.e.*, the relation of cause and effect is not the means of establishing it. This intends, 'I will mention *another* means of establishing it.'

c. [But some one may say,] Souls are nothing else than the body, and its organs, &c.: what need of imagining anything else? To this he replies:

Aph. 139.* Soul is something else than the body, &c.

a. [The meaning of the aphorism is] plain.

b. He propounds an argument in support of this:

Aph. 140.* Because that which is combined [and is, therefore, discerptible,] is for the sake of some other, [*not* discerptible].

a. That which is discerptible is intended for something else that is indiscerptible. If it were intended for something else that is discerptible, there would be a *regressus ad infinitum*.

b. And combinedness [involving (see § 67) discerptibleness,] consists in the Qualities' making some product by their state of mutual commixture; or [to express it otherwise,] combinedness is the state of the soft and the hard, [which distinguishes matter from spirit]. And this exists occultly in Nature, as well as the rest; because, otherwise, discerptibleness would not prove discoverable in the products thereof, *viz.*, the 'Great one,' &c.

c. He elucidates this same point:

Aph. 141.* [And Soul is something else than the body, &c.]; because there is [in Soul,] the reverse of the three Qualities, &c.

a. Because there is, in Soul, 'the reverse of the three Qualities,' &c., *i.e.*, because they are not seen [in it]. By the expression '&c.' is meant, because the *other* characters of Nature, also, are not seen [in soul].

b. He states another argument:

Aph. 142.* And [Soul is not material;] because of [its] superintedence [over Nature].

a. For a superintendent is an intelligent being; and Nature is unintelligent: such is the meaning.

b. He states another argument:

Aph. 143.* [And Soul is not material;] because of [its] being the experiencer.

a. It is Nature that is experienced; the experiencer is Soul. Although Soul, from its being unchangeably the same, is not [really] an experiencer, still the assertion [in the aphorism,] is made, because of the fact that the reflexion of the Intellect befalls it, [and thus makes it *seem* as if it experienced (see § 58. *a*.)].

b. Efforts are engaged in for the sake of Liberation. Pray, is this [for the benefit] of the Soul, or of Nature; [since Nature, in the shape of Mind, is, it seems, the experiencer]? To this be replies:

Aph. 144.* [It is for Soul, and not for Nature;] because the exertions are with a view to isolation [from all qualities; a condition to which Soul is competent, but Nature is not].

a. The very essence of Nature cannot depart from it [so as to leave it in the state of absolute, solitary isolation contemplated]; because the three Qualities are its very essence, [the departure of which from it would leave nothing behind], and because it would thus prove to be *not* eternal, [whereas, in reality, it *is* eternal]. The isolation (*kaivalya*) of that alone is possible of which the qualities are reflexional, [and not constitutive (see § 58. *a*.)]; and that is Soul.

b. Of what nature is this [Soul]? To this he replies:

Aph. 145.* Since light does not pertain to the unintelligent, light, [which must pertain to something or other, is the essence of the Soul, which, self-manifesting, manifests whatever else is manifest].

a. It is a settled point, that the unintelligent is not light; [it is not self-manifesting]. If Soul, also, were unintelligent [as the Naiyayikas hold it to be, in *substance*; knowledge being, by them, regarded not as its essence or substratum, but as one of its *qualities*], then there would need to be another light for *it*; and, as the simple theory, let Soul itself consist, essentially, of light.

b. And there is Scripture [in support of this view; for example, the two following texts from the *B[ihadaralyaka Upanishad*2]: 'Wherewith shall one distinguish that wherewith one distinguishes all this [world]?' 'Wherewith shall one take cognizance of the cognizer?'

c. [But the Naiyayika may urge,] *let* Soul be unintelligent [in its substance], but have Intelligence as its attribute. *Thereby* it manifests all things; but it is not, essentially, Intelligence. To this he replies:

Aph. 146.* It [Soul,] has not Intelligence as its attribute; because it is without quality.

a. If soul were associated with attributes, it would be [as we hold everything to be, that is associated with attributes,] liable to alteration; and, therefore, there would be no Liberation; [its attributes, or susceptibilities, always keeping it liable to be affected by something or other; or, the absolutely simple being the only unalterable].

b. He declares that there is a contradiction to Scripture in this, [*i.e.*, in the view which he is contending against]:

Aph. 147.* There is no denial [to be allowed] of what is established by Scripture; because the [supposed] evidence of intuition for this [*i.e.*, for the existence or qualities in the Soul,] is confuted [by the Scriptural declaration of the contrary].

a. The text, 'For this Soul is uncompanioned,' &c., would be confuted, if there were any annexation of qualities [to Soul: and the notion of confuting Scripture is not to be entertained for a moment].

b. But the literal meaning [of the aphorism] is this, that the fact, established by Scripture, of its [*i.e.*, soul's,] being devoid of qualities, &c., cannot be denied; because the Scripture itself confutes the [supposed] intuitive perception thereof, *i.e.*, the [supposed] intuitive perception of qualities, &c., [in the soul].

Aph. 148.* [If soul were unintelligent,] it would not be witness [of its own comfort,] in profound [and dreamless] sleep, &c.

a. If soul were unintelligent, then, in deep sleep, &c., it would not be a witness, a knower. But that this is not the case [may be inferred] from the phenomenon, that 'I slept *pleasanty*.' By the expression '&c.' [in the aphorism,] dreaming is included.

b. The Vedantis say that 'soul is *one* only'; and so, again, 'For Soul is eternal, omnipresent, changeless, void of blemish:' 'Being one [only], it is divided [into a seeming multitude] by Nature (*shakti*), *i.e.*, Illusion (*maya*), but not through its own essence, [to which there does not belong multiplicity].' In regard to this, he says [as follows]:

Aph. 149.* From the several allotment of birth, &c., a multiplicity of souls [is to be inferred].

a. 'Birth, &c.' By the '&c.,' growth, death, &c., are included. 'From the several allotment' of these, *i.e.*, from their being appointed; [birth to one, death to another, and so on]. 'A multiplicity of souls;' that is to say, souls are many. If soul were one only, then, when *one* is born, *all* must be born, &c.

b. He ponders, as a doubt, the opinion of the others, [*viz.*, of the Vedantis]:

Aph. 150.* [The Vedantis say, that,] there being a difference in its investments, moreover, multiplicity attaches [seemingly,] to the one [Soul]; as is the case with Space, by reason of jars, &c., [which mark out the spaces that they occupy].

a. As Space is one,—[and yet], in consequence of the difference of adjuncts, [as] jars, &c., when a jar is destroyed, it is [familiarly] said, 'the jar's space is destroyed' [for then there no longer exists a *space marked out by the jar*]; —so, also, on the hypothesis of there being but one Soul, since there is a difference of corporeal limitation, on the destruction thereof, [*i.e.*, of the limitation occasioned by any particular human body], it is merely a way of talking [to say], 'The soul has perished.' [This, indeed, is so far true, that there is really no perishing of Soul; but then it is true,] also on the hypothesis that there are *many* souls. [And it must be true:] otherwise, since Soul is eternal, [without beginning or end, as both parties agree], how could there be the appointment of birth and death?

b. He states [what may serve for] the removal of doubt [as to the point in question]:

Aph. 151.* The investment is different, [according to the Vedantis], but not that to which this belongs; [and the absurd consequences of such an opinion will be seen].

a. 'The investment is different,' [there are diverse bodies of John, Thomas, &c.]; 'that to which this belongs,' *i.e.* that [Soul] to which this investment [of body, in all its multiplicity,] belongs, is *not* different, [but is one only]: such is the meaning. And, [now consider], in consequence of the destruction of one thing, we are not to speak as if there were the destruction of something else; because this [if it were evidence of a thing's being destroyed,] would present itself where it ought not; [the destruction of Devadatta, *e.g.*, presenting itself, as a fact, when we are considering the case of Yajnadatta, who is not, for *that* reason, to be assumed to be dead]: and, on the hypothesis that Soul is one, the [fact that the Vedanta makes an] imputation of in consistent conditions is quite evident; since Bondage and Liberation do not [and cannot,] belong [simultaneously] to *one*. But the conjunction and [simultaneous] non-conjunction of the sky [or space] with smoke, &c., [of which the Vedanti may seek to avail himself, as an illustration,] are *not* contradictory; for Conjunction is not pervasion; [whereas, on the other hand, it would be nonsense to speak of Bondage as affecting one portion of a monad, and Liberation as affecting another portion; as a monkey may be in conjunction with a branch of a tree, without being in conjunction with the stem].

b. What may be [proved] by this? To this he replies:

Aph. 152.* Thus, [*i.e.*, by taking the Samkhya view,] there is no imputation of contradictory conditions to [a Soul supposed to be] everywhere present as *one* [infinitely extended monad].

a. 'Thus,' *i.e.*, [if you regard the matter rightly,] according to the manner here set forth, there is no 'imputation,' or attribution, 'of incompatible conditions,' Bondage, Liberation, &c., to a soul 'existing everywhere,' throughout all, as one, [*i.e.*, as a monad].

b. [But, the Vedanti may contend,] we *see* the condition of another attributed even to one quite different; as, *e.g.*, Nature's character as an agent [is attributed] to Soul, which is another [than Nature]. To this he replies:

Aph. 153.* Even though there be [imputed to Soul] the possession of the condition of another, this [*i.e.*, that it really possesses such,] is not established by the imputation; because it [Soul,] is *one* [absolutely simple, unqualified entity].

a. [The notion] that Soul is an agent is a mistake; because, that Soul is *not* an agent is true, and the imputation [of agency to Soul] is *not* true, and the combination of the true and the untrue is not real. Neither birth nor death or the like is compatible with Soul; because it is uncompanioned, [*i.e.*, unattended either by qualities or by actions].

b. [But the Vedanti may say:] and thus there will be an opposition to the Scripture. For, according to that, 'Brahma is one without a second:' 'There is nothing here diverse; death after death does he [deluded man,] obtain, who here sees, as it were, a multiplicity.' To this he replies:

Aph. 154.* There is no opposition to the Scriptures [declaratory] of the non-duality [of Soul]; because the reference [in such texts,] is to the *genus*, [or to Soul in general].

a. But there is no opposition [in our Samkhya view of the matter,] to the Scriptures [which speak] of the Oneness of Soul; because those [Scriptural texts] refer to the *genus*. By *genus* we mean sameness, the fact of being of the same nature: and it is to this alone that the texts about the non-duality [of Soul] have reference. It is not the indivisibleness [of Soul,—meaning, by its indivisibleness, the impossibility that there should be more souls than one,—that is meant in such texts]; because there is no motive [for viewing Soul as *thus* indivisible]: such is the meaning.

b. But then, [the Vedanti may rejoin,] Bondage and Liberation are just as incompatible in any single soul, on the theory of him who asserts that souls are many, [and that each is at once bound and free]. To this he replies:

Aph. 155.* Of him [*i.e.*, of that soul,] by whom the cause of Bondage is known, there is that condition [of isolation, or entire liberation], by the perception [of the fact, that Nature and soul are distinct, and that he, really, was *not* bound, even when he seemed to be so].

a. By whom is known 'the cause of bondage,' *viz*., the non-perception that Nature and soul are distinct, of him, 'by the perception' [of it], *i.e.*, by cognizing the distinction, there is 'that condition,' *viz*., the condition of isolation, [the condition (see § 144) after which the soul aspires. The soul in Bondage which is no real bondage may be typified by Don Quixote, hanging, in the dark, from the ledge of a supposed enormous precipice, and holding on for life, as he thought, from not knowing that his toes were within six inches of the ground].

b. [Well, rejoins the Vedanti,] Bondage [as you justly observe,] is dependent on non-perception [of the truth], and is not real. It is a maxim, that non-perception is removed by perception; and, on this showing, we recognize as correct the theory that soul is one, but not that of soul's being multitudinous. To this he replies:

Aph. 156.* No: because the blind do not see, can those who have their eyesight not perceive?

a. What! because a blind man does not see, does also one who has his eyesight not perceive? There are *many* arguments [in support of the view] of those who assert that souls are many, [though *you* do not see them]: such is the meaning.'

b. He declares, for the following reason, also, that souls are many:

Aph. 157.* Vamadeva, as well as others, has been liberated, [if we are to believe the Scriptures; therefore] non-duality is not [asserted, in the same Scriptures, in the Vedantic sense].

a. In the Puranas, &c., we hear, Vamadeva has been liberated,' 'shuka has been liberated,' and so on. If Soul were *one*, since the liberation of all would take place, on the liberation of one, the Scriptural mention of a diversity [of separate and successive liberations] would be self-contradictory.

b. [But the Vedanti may rejoin:] on the theory that Souls are many, since the world has been from eternity, and from time to time some one or other is liberated, so, by degrees, *all* having been liberated, there would be a universal void. But, on the theory that Soul is *one*, Liberation is merely the departure of an adjunct, [which, the Vedanti flatters himself, does not involve the inconsistency which he objects to the Samkhya], To this he replies:

Aph. 158.* Though it [the world,] has been from eternity, since, up to this day, there has not been [an entire emptying of the world], the future, also, [may be inferentially expected to be] thus [as it has been heretofore].

a. Though the world *has been* from eternity, since, up to this day, we have not seen it become a void, there is no proof [in support] of the view that there will be Liberation [of *all* Souls, so as to leave a void].

b. He states another solution [of the difficulty]:

Aph. 159.* As now [things are, so], everywhere [will they continue to go on: hence there will be] no absolute cutting short [of the course of mundane things].

a. Since souls are [in number,] without end, though Liberation successively take place, there will not be [as a necessary consequence,] a cutting short of the world. As now, so everywhere,—*i.e.*, in time to come, also,—there will be Liberation, but not, therefore, an absolute cutting short [of the world]; since of this the on-flowing is eternal.

b. On the theory, also, that Liberation is the departure of an adjunct [§ 157. *b*.], we should find a universal void; so that the doubt is alike, [in its application to either view]. Just as there might be an end of all things, on the successive liberation of many souls, so, since all adjuncts would cease, when [the fruit of] works [this fruit being in the shape of Soul's association with body, as its adjunct,] came to an end, the world would become void, [on the Vedanta theory, as well as on the Samkhya].

c. Now, [if the Vedanti says,] there will not be a void, because adjuncts are [in number,] endless, then it is the same, on the theory that Souls are many. And thus [it has been declared]: 'For this very reason, indeed, though those who are knowing [in regard to the fact that Nature and Soul are different], are continually being liberated, there will not be a void, inasmuch as there is no end of multitudes of souls in the universe.'

d. Pray, [some one may ask,] is Soul [*essentially*] bound? Or free? If [essentially] bound, then, since its essence cannot depart, there is no Liberation; for, if it [the essence,] departed, then it [Soul,] would [cease, with the cessation of its essence, and] not be eternal. If [on the other hand, you reply that it is essentially] free, then meditation and the like [which you prescribe for the attainment of liberation,] are unmeaning. To this he replies:

Aph. 160.* It [Soul,] is altogether free [but seemingly] multiform [or different, in appearance, from a free thing, through a delusive semblance of being bound].

a. It is not bound; nor is it liberated; but it is ever free, [see § 19]. But the destruction of ignorance [as to its actual freedom,] is effected by meditation, &c., [which are, therefore, not unmeaning, as alleged in § 159. *d*.].

b. It has been declared that Soul is a witness. Since it is a witness [some one may object], even when it has attained to discriminating [between Nature and Soul], there is no Liberation; [Soul, on this showing, being not an absolutely simple entity, but something *combined* with the character of a spectator or witness]. To this he replies:

Aph. 161.* It [Soul,] is a witness, through its connexion with sense-organs, [which quit it, on liberation].

a. A sense-organ is an organ of sense. Through its connexion therewith, it [Soul,] is a witness. And where is [its] connexion with sense-organs, [these products of Nature (see § 61)], when discrimination [between Nature and Soul] has taken place?

b. [Well, some one may ask], at all times of *what* nature is Soul? To this he replies:

Aph. 162.* [The nature of Soul is] constant freedom.

a. 'Constant freedom:' that is to say; Soul is, positively, always devoid of the Bondage called Pain [see §§ 1 and 19];

because Pain and the rest are modifcations of Understanding, [which (see § 61) is a modification of Nature, from which Soul is really distinct].

Aph. 163.* And, finally, [the nature of Soul is] indifference [to Pain and Pleasure, alike].

a. By 'indifference' is meant non-agency. The word *iti* [rendered 'finally,'] implies that the exposition of the Nature of Soul is completed.]

b. [Some one may say, the fact of] Soul's being an agent is declared in Scripture. How is this, [if, as you say, it be *not* an agent]? To this he replies:

Aph. 164.* [Soul's *fancy* of] being an agent is, through the influence [of Nature], from the proximity of Intellect, from the proximity of Intellect.

a. [Its] 'being an agent,' *i.e.*, Soul's fancy of being an agent, is 'from the proximity of Intellect,' 'through the influence' of Nature, [(see § 19,) of which Intellect (see § 61) is a modification].

b. The repetition of the expression 'from the proximity of Intellect' is meant to show that we have reached the conclusion: for thus do we see [practised] in the Scriptures, [*e.g.*, where it is said, in the Veda: 'Soul is to be known; it is to be discriminated from Nature: thus it does not come again, it does not come again'].

c. So much, in this Commentary on the illustrious Kapila's Aphorisms declaratory of the Samkhya, for the First Book, that on the [topics or] subject-matter [of the Samkhya system].

BOOK II

a. The subject-matter [of the Institute] has been set forth [in Book I.]. Now, in order to prove that it is not the *Soul* that undergoes the alterations [observable in the course of things], he will tell, very diffusely, in the Second Book, how the creation is formed out of the Primal Principle.

There, too, the nature of the products of Nature is to be declared fully, with a view to the very clear discrimination of Soul from these. Therefore, according to [the verses], *b*. 'Whoso rightly knows its changes, and the Primal Agent [Nature], and soul, the eternal, he, thirsting no more, is emancipated,'

c. we remark, that, with reference to the character, &c., of Emancipation, all the three [things mentioned in these verses] require to be known. And here, in the first place, with advertence to the consideration, that, if Nature, which is unintelligent, were to create without a motive, we should find even the emancipated one bound, he states the *motive* for the creation of the world:

Aph. 1.* Of Nature [the agency, or the being a maker, is] for the emancipation of what is [really, though not apparently,] emancipated, or else for [the removal of] itself.

a. The expression 'the being a maker' is borrowed from the last aphorism of the preceding Book. Nature makes the world for the sake of removing the pain, which is [really] a shadow [Book I., § 58], belonging to the Soul, which is, in its very nature, free from the bonds of pain; or [to explain it otherwise,] for the sake of removing pain [connected] by means of but a shadowy link; or [on the other hand,] it is 'for the sake of itself,' that is to say, for the sake of removing the actually real pain [which consists] of itself.

b. Although experience [of good and ill], also, as well as Emancipation, is a motive for creation, yet Emancipation alone is mentioned, inasmuch as it is the principal one.

c. But then, if creation were for the sake of Emancipation, then, since Emancipation might take place through creation once for all, there would not be creation again and again; to which he replies:

Aph. 2.* Because this [Emancipation] is [only] of him that is void of passion.

a. Emancipation does not take place through creation once for all; but it is [the lot only] of him that has been extremely tormented many times by the various pain of birth, death, sickness, &c.; and, therefore, [successive creation goes on] because Emancipation actually occurs in the case only of him in whom complete dispassion has arisen through the knowledge of the distinctness of Nature and Soul: such is the meaning.

b. He tells the reason why dispassion does not take place through creation once for all:

Aph. 3.* It is not effected by the mere hearing; because of the forcibleness of the impressions from eternity.

a. Even the hearing [of Scripture, in which the distinctness of Nature from Soul is enounced,] comes [not to all alike, but only] through the merit of acts done in many births, [or successive lives]. Even then dispassion is not established through the mere hearing, but through direct cognition; and direct cognition does not take place suddenly, because of the forcibleness of false impressions that have existed from eternity, but [the required direct cognition takes place] through the completion of Concentration; and there is an abundance of obstacles to Concentration [see Yoga Aphorisms, Book II]: therefore, only after many births do dispassion and Emancipation take place at any time of any one at all: such is the meaning.

b. He states another reason for the continuous flow of creation:

Aph. 4.* Or as people have, severally, many dependants.

a. As householders have, severally, many who are dependent upon them, according to the distinctions of wife, children, &c., so, also, the Qualities, *viz*., Goodness, &c., [Book I., § 61. *b*.] have to emancipate innumerable Souls, severally. Therefore, however many Souls may have been emancipated, the onflow of creation takes place for the emancipation of other Souls; for Souls are [in number,] without end: such is the meaning. And so the Yoga aphorism [Book II., § 22] says: 'Though it have ceased to be, in respect of him that has done the work, it has not [absolutely] ceased to be; because it is common to others besides him.'

b. But then why is it asserted that Nature alone creates, when, by the text, 'From that or this Soul proceeded the Ether,' &c., it is proved that *Soul*, also, creates? To this he replies:

Aph. 5.* And, since it [the character of creator,] belongs, really, to Nature, it follows that it is fictitiously attributed to Soul.

a. And, since Nature's character of creator is decided to be real, there is, really, in the Scriptures, only a fictitious [or figurative] attribution of creativeness to Soul.

b. But then, if it be thus, how is it laid down that Nature's creativeness, moreover, is *real*; since we are told [in Scripture,] that creation, moreover, is on a level with a dream? To this he replies:

Aph. 6.* Since it is proved from the products.

a. That is to say: because the real creative character of Nature is established just 'from the products,' *viz*., by that evidence [see Book I., § 110,] which acquaints us with the subject [in which the creative character inheres]; for *products* are real, inasmuch as they produce impressions and exhibit acts. [The reality of eternal things is established here, just as it is by Locke, who says: 'I think God has given me assurance enough as to the existence of things without me; since, by their different application, I can produce, in myself, both *pleasure and pain* (*artha*), which is one great concernment of my present state.' These existing products being admitted, the Samkhya argues that they must have a cause; and, as this cause means neither more nor less than something creative, whatever proves the existence of the cause proves, at the same time, its creative character.]

b. But then [it may be said], on the alternative [see § 1] that Nature works for *herself*, she must energize with reference to the *emancipated* Soul, also. To this he replies:

Aph. 7.* The rule is with reference to one knowing; just as escape from a thorn.

a. The word *chetana* here means 'one knowing;' because the derivation is from *chit*, 'to be conscious'. As one and the same thorn is not a cause of pain to him who, being 'one knowing,' *i.e.*, aware of it, escapes from that same, but actually is so in respect of *others*; so Nature, also, is escaped by 'one knowing,' one aware, one who has accomplished the matter: to *him* it does not consist of pain; but to others, who are *not* knowing, it actually is a cause of pain: such is the 'rule,' meaning, the distribution. Hence, also, of Nature, which is, by its own nature, bound [inasmuch as it *consists* of bonds], the self-emancipation is possible; so that it does not energize with reference to the emancipated Soul [§ 6. *b*.].

b. But then [suggests some one], what was said [at § 5], that, in respect of Soul, the creative character is only fictitiously attributed, this is not proper; because it is fitting, that, by the conjunction of Nature, *Soul*, also, should be modified into Mind, &c.; for a modification of wood, &c., resembling earth, &c., through the conjunction of earth, &c., is seen: to which he replies:

Aph. 8.* Even though there be conjunction [of Soul] with the other [*viz*., Nature], this [power of giving rise to products] does not exist in it immediately; just like the burning action of iron.

a. Even though there be conjunction with Nature, there belongs to Soul no creativeness, 'immediately,' *i.e.*, directly. An illustration of this is, 'like the burning action of iron:' as iron does not possess, directly, a burning Power; but this is only fictitiously attributed to it, being through the fire conjoined with it: such is the meaning. But, in the example just mentioned, it is admitted that there is an alteration of both; for this is proved by sense-evidence: but, in the instance under doubt, since the case is accounted for by the modification of one only, there is cumbrousness in postulating the modification of both; because, otherwise, by the conjunction of the China-rose, it might be held that the colour of the crystal was changed.

b. It has already been stated [§ 1] that the fruit of creation is emancipation. Now he states the principal occasional cause of creation:

Aph. 9.* When there is passion, or dispassion, there is concentration, [in the latter case, and] creation, [in the former].

a. When there is passion, there is creation; and, when there is dispassion, there is 'concentration,' *i.e.*, the abiding [of Soul] in its own nature [see Yoga Aphorisms, Book I., § 32]; in short, emancipation, or the hindering of the modifications of the thinking principle [Yoga Aphorisms, Book I., § 22]: such is the meaning. And so the import is, that Passion is the cause of creation; because of their being3 simultaneously present or absent.

b. After this he begins to state the manner of creation:

Aph. 10.* In the order [see § 12. *b*.] of Mind, &c., [is the creation] of the five elements, [or of the material world].

a. 'Creation' is supplied from the preceding aphorism.

b. He mentions a distinction [between these successively creative energies and the primal one]:

Aph. 11.* Since creation is for the sake of Soul the origination of these [products of Nature] is not for their own sake.

a. 'Of these,' *i.e.*, of Mind, &c., since the creativeness is 'for the sake of Soul,' *i.e.*, for the sake of the emancipation of Soul, the 'origination,' *i.e.*, the creativeness, is not for the sake of themselves; since, inasmuch as they are perishable, they [unlike Nature, (see § 1)] are not susceptible of emancipation: such is the meaning.

b. He declares the creation of limited space and time:

Aph. 12.* [Relative] Space and Time [arise] from the Ether, &c.

a. The Space and Time which are *eternal* [and absolute], being the *source* of the Ether, are, really, sorts of qualities of Nature: therefore it is consistent that Space and Time should be all-pervading. But the Space and Time which are limited arise from the Ether, through the conjunction of this or that limiting object: such is the meaning. By the expression '&c.,' [in the aphorism,] is meant 'from the apprehending of this or that limiting object.'

b. Now he exhibits, in their order, through their nature and their habits, the things mentioned [in § 10] as 'in the order of Mind, &c.':

Aph. 13.* Intellect is judgment.

a. 'Intellect' is a synonym of 'the Great Principle' [or Mind (see Book I., § 71)]; and 'judgment,' called [also] ascertainment, is its peculiar modification: such is the meaning. But they are set forth as identical, because a property and that of which it is the property are indivisible. And it is to be understood, that this Intellect is 'Great,' because it pervades all effects other than itself, and because it is of great power.

b. He mentions other properties, also, of the Great Principle:

Aph. 14.* Merit, &c., are products of it.

a. The meaning is, that Merit, Knowledge, Dispassion, and Supernatural Power, moreover, are formed out of *intellect*, not formed of *self-consciousness* (*ahankara*), &c.; because intellect alone [and not self-consciousness,] is a product of superlative Purity, [without admixture of Passion and Darkness].

b. But then, if it be thus, how can the prevalence of demerit, in the portions of intellect lodged in men, cattle, &c., be accounted for? To this he replies:

Aph. 15.* The Great one [intellect,] becomes reversed through tincture.

a. That same 'Great one,' *i.e.*, the Great Principle [or intellect], through being tinged with Passion and Darkness, also becomes 'reversed' [see § 14. *a*.], *i.e.*, vile, with the properties of Demerit, Ignorance, Non-dispassion, and want of Supernatural Power: such is the meaning.

b. Having characterized the Great Principle, he defines its product, Self-consciousness:

Aph. 16.* Self-consciousness is a conceit.

a. 'Self-consciousness' is what makes the Ego, as a potter [makes a pot]; the thing [called] the internal* instrument (*anta-karana*): and this, inasmuch as a property and that of which it is the property are indivisible, is spoken of as 'a conceit,' [*viz.*, of personality], in order to acquaint us that this is its peculiar modification. Only when a thing has been determined by intellect [*i.e.*, by an act of judgment (see § 13. *a*.)], do the making of an Ego and the making of a Meum take place.

b. He mentions the product of Self-consciousness, which has arrived in order:

Aph. 17.* The product of it [*viz.*, of Self-consciousness,] is the eleven [organs], and the five Subtile Elements.

a. The meaning is, that the eleven organs, with the five Subtile Elements, *viz.*, Sound, &c., are the product of Self-consciousness.

b. Among these, moreover, he mentions a distinction:

Aph. 18.* The eleventh, consisting of [the principle of] Purity, proceeds from modified Self-consciousness.

a. The 'eleventh,' *i.e.*, the completer of the eleven, *viz.*, Mind, [or the 'internal organ,'—which is not to be confounded with 'the Great one,' called also Intellect and Mind,—alone,] among the set consisting of sixteen [§ 17], consists of Purity; therefore it is produced from Self-consciousness 'modified,' *i.e.*, pure: such is the meaning. And hence, too, it is to be reckoned that the ten organs are from the Passionate Self-consciousness; and the Subtile Elements, from the Dark Self-consciousness.

b. He exhibits the eleven organs:

Aph. 19.* Along with the organs of action and the organs of understanding another is the eleventh.

a. The organs of action are five, *viz.*, the vocal organ, the hands, the feet, the anus, and the generative organ; and, the organs of understanding are five, those called the organs of sight, hearing, touch, taste, and smell. Along with these ten, 'another,' *viz.*, Mind, is 'the eleventh,' *i.e.*, is the eleventh organ: such is the meaning.

b. He refutes the opinion that the Organs are formed of the Elements:

Aph. 20.* They [the organs,] are not formed of the Elements; because there is Scripture for [their] being formed of Self-consciousness.

a. Supply 'the organs.'

b. Pondering a doubt, he says:

Aph. 21.* The Scripture regarding absorption into deities is not [declaratory] of an originator.

a. That Scripture which there is about absorption into

deities is not 'of an originator,' that is to say, it does not refer to an originator; because [although a thing, *e.g.*, a jar, when it ceases to be a jar, is usually spoken of as being resolved into its originator, *viz.*, into earth, yet] we see the absorption of a drop of water into what, nevertheless, is *not* its originator, *viz.*, the ground; [and such is the absorption into a deity from whom the Mind absorbed did not originally emanate].

b. Some say that the Mind, included among the organs, is eternal. He repels this:

Aph. 22.* [None of the organs is eternal, as some hold the Mind to be;] because we have Scripture for their beginning to be, and because we see their destruction.

a. All these organs, without exception, have a beginning; for the Scripture says, 'From this are produced the vital air, the mind, and all the organs;' &c., and because we are certified of their destruction by the fact that, in the conditions of being aged, &c., the mind, also, like the sight and the rest, decays, &c.: such is the meaning.

b. He rebuts the atheistical opinion that the sense [for example,] is merely the set of eye-balls, [&c.]:

Aph. 23.* The Sense is supersensuous; [it being the notion] of mistaken persons [that the Sense exists] in [identity with] its site.

a. Every Sense is supersensuous, and not perceptible; but only in the opinion of mistaken persons does the Sense exist 'in its site,' *e.g.*, [Sight,] in the eye-ball, in the condition of identity [with the eye-ball]: such is the meaning. The correct reading is: '[The sense is something supersensuous; to confound it with] the site, [is a mistake].'

b. He rebuts the opinion that one single Sense, through diversity of powers, performs various offices:

Aph. 24.* Moreover, a difference being established if a difference of *powers* be [conceded], there is not a oneness [of the organs].

a. Even by the admission that a diversity of *powers* belongs to one single organ, the diversity of organs is established;

because the *powers* are, assuredly, organs; therefore, there is not a singleness of organ: such is the meaning.

b. But then [it may be said], there is something unphilosophical in supposing various kinds of organs to arise from one single Self-consciousness. To this he replies:

Aph. 25.* A theoretical discordance is not [of any weight,] in the case of what is matter of ocular evidence.

a. This is simple.

b. He tells us that, of the single leading organ, the Mind, the other ten are kinds of powers:

Aph. 26.* The Mind identifies itself with both

a. That is to say: the Mind identifies itself with the organs of intellection and of action.

b. Of his own accord, he explains the meaning of the expression 'identifies itself with both:'

Aph. 27.* By reason of the varieties of transformation of [which] the Qualities [are susceptible], there is a diversity [of their product, the Mind,] according to circumstances.

a. As one single man supports a variety of characters, through the force of association,—being, through association with his beloved, a lover; through association with one indifferent, indifferent; and, through association with some other, something other,—so the Mind, also, through association with the organ of vision, or any other, becomes various, from its becoming one with the organ of vision, or any other; by its being [thereby] distinguished by the modification of seeing, or the like. The argument in support of this is, 'of the Qualities,' &c.; the meaning being, because of the adaptability of the Qualities, Goodness, &c., to varieties of transformation.

b. He mentions the object of the organs of intellection and of action:

Aph. 28.* Of both [sets of organs the object is that list of things], beginning with Colour, and ending with the dirt of Taste.

a.. The 'dirt' of the tastes of food, &c., means ordure, &c.,

[into which the food, consisting of the *quality* Taste, &c., is partly transformed].

b. Of what Soul (*indra*), through what service, these are termed Organs (*indriya*), both these things he tells us:

Aph. 29.* The being the seer, &c., belongs to the Soul; the instrumentality belongs to the Organs.

a. For, as a king, even without himself energizing, becomes a warrior through his instrument, his army, by directing this by orders simply, so the Soul, though quiescent, through all the organs, of vision, &c., becomes a seer, a speaker, and a judger, and the like, merely through the proximity called 'Conjunction;' because it moves these, as the lodestone [does the iron, without exerting any effort].

b. Now he mentions the special modifications of the triad of internal organs:

Aph. 30.* Of the three [internal organs] there is a diversity among themselves.

a. The aspect of Intellect is attention; of Self-consciousness, conceit [of personality]; of the Mind, decision and doubt.

b. He mentions, also, a common aspect of the three:

Aph. 31.* The five airs, *viz.*, Breath, &c., are the modifications, in common, of the [three internal] instruments.

a. That is to say: the five, in the shape of Breath, &c., which are familiarly known as 'airs', because of their circulating as the air does, these [animal spirits] are the joint or common 'modification,' or kinds of altered form, 'of the instruments,' *i.e.*, of the triad of internal instruments.

b. The opinion is not ours, as it is that of the Vaisheshikas, that the modifications of the organs take place successively only, and not simultaneously. So he says:

Aph. 32.* The modifications of the organs take place successively and simultaneously.

a. This is simple.

b. Lumping the modifications of the understanding, with a view to showing how they are the cause of the world, he, in the first place, exhibits [them]:

Aph. 33.* The modifications [of the understanding, which are to be shown to be the cause of the world, and] which are of five kinds, are [some of them,] painful and [others,] not painful.

a. That the modifications are of five sorts is declared by Patanjali's aphorism, [see Yoga Aphorisms, Book I., § 6]

b. He acquaints [us] with the nature of Soul:

Aph. 34.* On the cessation thereof [*viz.*, of mundane influences], its tincture ceasing, it [Soul,] abides in itself.

a. That is to say: during the state of repose of these modifications, it [the Soul], the reflexion of these having ceased, is abiding in itself; being, at *other* times, also, as it were, in isolation, [though seemingly not so]. And to this effect there is a triad of Aphorisms of the Yoga, [*viz.*, Book I., §§ 2, 3, and 4].

b. He explains this by an illustration:

Aph. 35.* And as [by] a flower, the gem.

a. The 'and' implies that this is the *reason* [of what was asserted in the preceding aphorism]; the meaning being, as the gem [is tinged, apparently,] by a flower. As the gem called rock-crystal, by reason of a flower of the Hibiscus, becomes red, not abiding in its own state, and, on the removal thereof, becomes colourless, abiding in its own state, in like manner [is the Soul apparently tinged by the adjunction of the Qualities].

b. But then [it may be asked], by whose *effort* does the aggregate of the organs come into operation; since Soul is motionless, and since it is denied that there is any Lord [or Demiurgus]? To this he replies:

Aph. 36.* The Organs also arise, for the sake of Soul from the development of *desert*.

a. The meaning is, that, just as Nature energizes 'for the sake of Soul,' so 'the Organs also arise;' *i.e.*, the energizing of the Organs is just in consequence of the development of the deserts of the soul: [see Yoga Aphorisms, Book II., § 13. *b.*]. And the desert belongs entirely to the investment; [the Soul not really possessing either merit or demerit].

b. He mentions an instance of a thing's spontaneously energizing for the sake of another:

Aph. 37.* As the cow for the calf.

a. As the cow, for the sake of the calf, quite spontaneously secretes milk, and awaits no other effort, just so, for the sake of the master, Soul, the Organs energize quite spontaneously: such is the meaning. And it is *seen*, that, out of profound sleep, the understanding of its own accord wakes up.

b. With reference to the question, how many Organs there are, external and internal combined, he says:

Aph. 38.* Organ is of thirteen sorts, through division of the subordinates.

a. The triad of internal organs, and the ten external organs, combined, are thirteen. He says 'sorts,' in order to declare that, of these, moreover, there is an infinity, through [their] distinction into individuals. He says 'through division of the subordinates,' with a reference to the fact, that it is *understanding* which is the *principal* organ; the meaning being, because the organs [or functions,] of the single organ, called understanding, are more than one.

b. But then, since understanding [it seems,] alone is the principal instrument in furnishing its object [of emancipation] to Soul, and the instrumentality of the others is secondary, in this case what is [meant by] *secondariness*? [Why are they said to be instrumental *at all*?] In regard to this he says:

Aph. 39.* Because the quality of being most efficient is conjoined with the organs; as in the case of an axe.

a. The quality of the [principal] organ, the understanding, in the shape of being most efficient on behalf of soul, exists, derivatively, in the [other derivative] organs. Therefore it is made out that an organ is of thirteen kinds: such is the connexion with the preceding aphorism:

b. 'As in the case of an axe.' As, although the blow itself, since it is this that puts an end to our non-possession of the result, is the principal efficient in the cutting, yet the axe, also, is an efficient, because of its close proximity to the quality of being the principal efficient, so [here, also]: such is the meaning. He does not here say that Self-consciousness is secondarily efficient, meaning to imply that it is one with the internal organ.*

c. Specifying the precise state of the case in regard to the condition of secondary and principal, he says:

Aph. 40.* Among the two [the external and the internal organs], the principal is Mind; just as, in the world, among troops of dependants.

a. 'Among the two,' *viz*., the external and the internal, 'Mind,' *i.e*., understanding, simply, is 'the principal,' *i.e*., chief; in short, is the immediate cause; because it is that which furnishes Soul with its end; just as, among troops of dependants, some one single person is the prime minister of the king; and the others, governors of towns, &c., are his subordinates: such is the meaning.

b. Here the word 'Mind' does not mean the third internal organ, [(§ 30. *a*.) but Intellect, or 'the Great One.']

c. He tells, in three aphorisms, the reasons why Intellect [or understanding] is the principal:

Aph. 41.* [And Intellect is the principal, or immediate and direct, efficient in Soul's emancipation;] because there is no wandering away.

a. That is to say: because it [understanding,] pervades all the organs; or because there is no result apart from it.

Aph. 42.* So, too, because it [the understanding,] is the depository of all self-continuant impressions.

a. Understanding alone is the depository of all self-continuant impressions, and not the Sight, &c., or Self-conconsciousness, or the Mind; else it could not happen that things formerly seen, and heard, &c., would be remembered by the blind, and deaf, &c.

Aph. 43.* And because we infer this [its preeminence] by reason of its meditating.

a. That is to say: and because we infer its preeminence, 'by reason of its meditating,' *i.e*., its modification in the shape of meditation. For the modification of thought called 'meditation' is the noblest of all the modifications [incident to Soul, or pure Thought, whose blessedness, or state of emancipation, it is to have no *modification* at all]; and the Understanding itself which, as being the depository thereof, is, further, named Thought [*chitta*, from the same root as *chinta*], is nobler than the organs whose modifications are other than this: such is the meaning.

b. But then, suppose that the modification 'meditation' belongs only to the *Soul*, [suggests some one]. To this he replies:

Aph. 44.* It cannot be of its own nature.

a. That is to say: meditation cannot belong to Soul essentially; because of the immobility [of Soul; whereas 'meditation' is an effort].

b. But then, if thus the preeminence belongs to understanding alone, how was it said before [at § 26,] that it is the *Mind* that takes the nature of both [sets of organs, in apparent contradiction to the view propounded at § 39]? To this he replies:

Aph. 45.* The condition [as regards Soul's instruments,] of secondary and principal is relative; because of the difference of function.

a. In respect to the difference of function, the condition, as secondary, or principal, of the instruments [of Soul] is relative. In the operations of the Sight, &c., the Mind is principal; and, in the operation of the Mind, Self-consciousness, and, in the operation or Self-consciousness, Intellect, is principal [or precedent].

b. But then, what is the cause of this arrangement; *viz*., that, of this [or that] Soul, this [or that] Intellect, alone, and not another Intellect, is the instrument? With reference to this, he says:

Aph. 46.* The energizing [of this or that Intellect] is for the sake of this [or that Soul]; because of [its] having been purchased by the works [or deserts] of this [or that Soul]; just as in the world.

a. The meaning is, that, 'the energizing,' *i.e.*, all operation, of the instrument is for the sake of this [or that] Soul; because of [its] having been purchased by this [or that] Soul's works [or deserts]; just as in the world. As, in the world [or in ordinary affairs], whatever axe, or the like, has been purchased by the act, *e.g.*, of buying, by whatever man, the operation of that [axe, or the like], such as cleaving, is only for the sake of that man [who purchased it]: such is the meaning. The import is, that *therefrom* is the distributive allotment of instruments [inquired about under § 45. *b*.]

b. Although there is *no* act in Soul, because it is immovable, still, since it is the means of Soul's experience, it is *called* the act of Soul; just like the victories, &c., of a king [which are, really, the acts of his servants]; because of Soul's being the owner [of the results of acts; as the king is of the results of the actions of his troops].

c. In order to make clear the chiefship of Intellect, he sums up, [as follows]:

Aph. 47.* Admitting that they [the various instruments of Soul, all] equally act, the preeminence belongs to Intellect; just as in the world, just as in the world.

a. Although the action of all the instruments is the same, in being for the sake of Soul, still the preeminence belongs to Intellect alone: just as in the world. The meaning is, because it is just as the preeminence, in the world, belongs to the prime minister, among the rulers of towns, and the rest, even although there be no difference so far as regards their being [all alike workers] for the sake of the king. Therefore, in all the Institutes, Intellect alone is celebrated as 'the Great One.' The repetition [*viz.*, 'just as in the world, just as in the world,'] implies the completion of the Book.

b. So much for [this abstract of] the Second Book, on the Products of Nature, in the commentary, on Kapila's Declaration of the Samkhya, composed by the venerable Vijnana Acharya.

BOOK III

a. In the next place, the gross product of Nature, *viz.*, the great elements and the dyad of bodies, is to be described; and, after that, the going into various wombs, and the like; [this description being given] with a view to that less perfect degree of dispassionateness which is the cause of one's engaging upon the means of knowledge; and, after that, with a view to perfect freedom from passion, all the means of knowledge are to be told: so the Third [Book] commences:

Aph. 1.* The origination of the diversified [world of sense] is from that which has no difference.

a. '[Which] has no difference,' *i.e.*, that in which there exists not a distinction, in the shape of calmness, fierceness, dulness, &c., *viz.*, the Subtile Elements, called 'the five somethings, simply;' from this [set of five] is the origination of 'the diversified,' [so called] from their possessing a difference, in the shape of the calm, &c., *viz.*, the gross, the great Elements: such is the meaning. For, the fact of consisting of pleasure, or the like, in the shape of the calm, and the rest, is manifested, in the degrees of greater, and less, &c., in the gross Elements only, not in the Subtile; because *these*, since they have but the one form of the calm, are manifest to the concentrated, [practitioners of meditation, but to no others].

b. So then, having stated, by composing the preceding Book, the origin of the twenty-three Principles, he states the origination, therefrom, of the dyad of bodies:

Aph. 2.* Therefrom, of the Body.

a. 'Therefrom,' *i.e.*, from the twenty-three Principles, there is the origination of the pair of Bodies, the Gross [Body] and the Subtile: such is the meaning.

b. Now he proves that mundane existence could not be accounted for otherwise than on the ground of the twenty-three Principles.

Aph. 3.* From the seed thereof is mundane existence.

a. 'Thereof,' *i.e.*, of the Body; 'from the seed,' *i.e.*, from the Subtile one, as its cause, in the shape of the twenty-three Principles, is 'mundane existence,' *i.e.*, do the going and coming of Soul take place; for it is impossible that, of itself, there should be a going, &c., of that which, in virtue of [its] all-pervadingness, is immovable: such is the meaning. For Soul, being conditioned by the twenty-three Principles, only by means of that investment migrates from Body to Body, with a view to experiencing the fruits of previous works.

b. He states, also, the limit of mundane existence.

Aph. 4.* And, till there is discrimination, there is the energizing of these, which have no differences.

a. The meaning is, that, of all Souls whatever, void of the differences of being Lord, or not Lord, &c., [though, seemingly, possessed of such differences,] 'energizing,' *i.e.*, mundane existence, is inevitable, even till there is discrimination [of Soul from its seeming investments]; and it does not continue after that.

b. He states the reason of this:

Aph. 5.* Because of [the necessity of] the other's experiencing.

a. The meaning is: because of the necessity that the other, *i.e.*, that that very [Soul], which does not discriminate, should experience the fruit of its own [reputed] acts.

b. He states, that, even while there is a Body, during the time of mundane existence, fruition [really] is not:

Aph. 6.* It [Soul,] is now quite free from both.

a. 'Now,' *i.e.*, during the time of mundane existence. Soul is quite free 'from both,' *i.e.*, from the pairs, *viz.*, cold and heat, pleasure and pain, &c.: such is the meaning.

b. He next proceeds to describe, separately, the dyad of Bodies:

Aph. 7.* The Gross [Body] usually arises from father and mother; the other one is not so.

a. The Gross one arises from father and mother, 'usually,' *i.e.*, for the most part; for there is mention also of a Gross Body *not* born of a womb: and 'the other,' *i.e.*, the Subtile Body, is 'not so,' *i.e.*, does not arise from a father and mother; because *it* arises from creation, &c.: such is the meaning.

b. He decides [the question], through disguise by which one of the Bodies, Gross and Subtile, the conjunction of the pairs [pleasure and pain, &c.,] with Soul takes place.

Aph. 8.* To that which arose antecedently it belongs to be that whose result is this; because it is to the one that there belongs fruition, not to the other.

a. 'To be that whose result is this,' *i.e.*, to have pleasure and pain as its effect [reflected in Soul], belongs to that Subtile Body alone whose origin was 'antecedent,' *i.e.*, at the commencement of the creation [or annus magnus]. Why? Because the fruition of what is called pleasure and pain belongs only to 'the one,' *i.e.*, the Subtile Body, but not to 'the other,' *i.e.*, the Gross Body; because all are agreed that there is neither pleasure nor pain, &c., in a body of *earth*: such is the meaning.

b. He tells the nature of the Subtile Body just mentioned:

Aph. 9.* The seventeen, as one, are the Subtile Body.

a. The Subtile Body, further, through its being container and contained, is twofold. Here the seventeen, [presently mentioned,] mingled, are the Subtile Body; and that, at the beginning of a creation, is but one, in the shape of an aggregate; [as the forest, the aggregate of many trees, is but one]: such is the meaning. The seventeen are the eleven organs, the five Subtile Elements, and Understanding. Self-consciousness is included under Understanding.

b. But [one may ask,] if the Subtile Body be *one*, how can there be diverse experiences accordingly as Souls are [numerically] distinct, [one from another]? To this he replies:

Aph. 10.* There is distinction of individuals, through diversity of desert.

a. Although, at the beginmng of the creation [or annus magnus], there was but one Subtile Body, in the shape of that investment [of Soul (see *Vedanta-sara*, § 62.) named] *Hiranyagarbha*, still, subsequently, moreover, there becomes a division of it into individuals,—a plurality, partitively, in the shape of individuals;—as, at present, there is, of the one Subtile Body of a father, a plurality, partitively, in the shape of the Subtile Body of son, daughter, &c. He tells the cause of this, saying, 'through diversity of desert;' meaning, through actions, &c., which are causes of the experiences of other animal souls.

b. But then, on this showing, since the Subtile one alone, from its being the site of fruition, is [what ought to be denoted by the term] *Body*, how is the term Body applied to the Gross one? To this he replies:

Aph. 11.* From its being applied to *it*, [*viz*., to the Subtile one], it is applied to the Body, which is the tabernacle of the abiding thereof.

a. But then, what proof is there of another body,—other than the one consisting of the six sheaths,—serving as a tabernacle for the Subtile Body? With reference to this, he says:

Aph. 12.* Not independently [can the Subtile Body exist], without that [Gross Body]; just like a shadow and a picture.

a. That is to say: the Subtile Body does not stand independently, 'without that,' *i.e.*, without a support; as a shadow, or as a picture, does not stand without a support. And so, having abandoned a Gross Body, in order to go to another world, it is settled that the Subtile Body takes another body, to serve as its tabernacle: such is the import.

b. But then [it may be said], of the Subtile Body, since it is limited substance, as the Air, or the like, let the *Ether* [or *Space*], without [its] being attached [to anything], be the site: it is purposeless to suppose [its] attachment to anything else. To this he replies:

Aph. 13.* No, even though it be limited; because of [its] association with masses; just like the sun.

a. Though it be limited, it does not abide independently, without association; for, since, just like the sun, it consists of *light*, it is inferred to be associated with a mass: such is the meaning, All lights, the sun and the rest, are seen only under the circumstances of association [of the luminiferous imponderable] with earthy subtances; and the Subtile Body consists of 'Purity,' which is Light: therefore it must be associated with the Elements.

b. He determines the magnitude of the Subtile Body:

Aph. 14.* It is of atomic magnitude; for there is a Scripture for its acting.

a. 'It,' the Subtile Body, is 'of atomic magnitude,' *i.e.*, limited, but not absolutely an *atom*; because it is declared to have parts. Wherefore? 'For there is Scripture for its acting;' *i.e.*, because there is Scripture about its acting. When a thing is all-prevading, it cannot act; [action being motion]. But the proper reading is, 'because there is Scripture for its *moving*.'

b. He states another argument for its being limited:

Aph. 15.* And because there is Scripture for its being formed of food.

a. That is to say: it, *viz.*, the Subtile Body, cannot be all-pervading; because there is a Scripture for its being partially formed of food; for, if it were all-pervading, it would be eternal. Although Mind, &c., are not formed of the Elements, still it is to be understood that they are spoken of as formed of food, &c.; because they are filled with homogeneous particles, through contact with food; [as the light of a lamp is supplied by contact with the oil].

b. For what purpose is the mundane existence, the migrating from one body to another [Gross] body, of Subtile Bodies, which are *unintelligent*? With reference to this, he says:

Aph. 16.* The mundane existence of is for the sake of Soul; just like a king's cooks.

a. That is to say: as the cooks of a king frequent the kitchens for the sake of the king, so the Subtile Bodies transmigrate for the sake of Soul.

b. The Subtile Body has been discussed in respect of all its peculiarities. He now likewise discusses the Gross Body, also:

Aph. 17.* The Body consists of the five elements.

a. That is to say: the Body is a modification of the five elements mingled.

b. He mentions another opinion:

Aph. 18.* Some say it consists of *four* elements.

a. This [is alleged] with the import that the *Ether* does not originate [anything].

Aph. 19.* Others say that it consists of *one* element.

a. The import is, that the body is of Earth only, and the other elements are merely supporters. Or 'of one element' means, of one or other element: [see the Rosicrucian doctrine in the *Tarka-sangraha*, § 13., &c].

b. He tells us what is proved by the fact that the Body consists of the Elements:

Aph. 20.* Intellect is not natural [a natural result of organization]; because it is not found in them severally.

a. That is to say: since we do not find intellect in the separated Elements, intellect is not natural to the Body,—which consiste of the Elements,—but is adventitious.

b. He states another refutation [of the notion that Intellect is a property of the Body]:

Aph. 21.* And [if the Body had intellect natural to it,] there would not be the death, &c., of anything.

a. That is to say: and, if the Body had intellect natural to it, there would not be the death, the profound sleep, &c., 'of anything,' of all things. For death, profound sleep, &c., imply the body's being non-intelligent; and this, if it were, by its own nature, intelligent, would not take place; because the essential nature of a thing remains as long as the thing remains.

b. Pondering a doubt, as to the assertion [in § 20], *viz.*, 'because it is not found in them severally,' he repels it:

Aph. 22.* If [you say that Intellect results from organization, and that] it is like the power of something intoxicating, [the ingredients of which, separately, have no intoxicating power, we reply, that] this might arise, on conjunction, if we had seen, in each [element, something conducive to the result].

a. But then, as an intoxicating power, though not residing in the substances severally, resides in the mixed substance, so may Intellect, also, be; if any one say this, it is not so. If it *had been* seen in each [constituent], its appearance in the compound might have had place; but, in the case in question, it is not the case that it is seen in each. Therefore, in the illustration [of something intoxicating resulting from mixture], it being established, by the Institutes, &c., that there is, in each ingredient, a *subtile* tendency to intoxicate, it is settled only that, at the time when these combine, there will be a *manifestation* of the [latent] power of intoxicating; but, in the thing illustrated, it is not established, by any proof whatsoever, that there is intelligence, in a subtile [or undeveloped] state, in the elements separately: such is the meaning.

b. It was stated [§ 16,] that the Subtile Bodies transmigrate for the sake of Soul. In regard to this, he tells, in two aphorisms, by what operation, dependent on the birth of the Subtile Bodies, which means their transmigrations into Gross Bodies, what aims of Soul are accomplished:

Aph. 23.* From knowledge [acquired during mundane existence, comes] salvation, [Soul's *chief* end].

a. That is to say: by the transmigration of the Subtile Body, through birth, there takes place the direct operation of discrimination [between Soul and Non-soul]; [and] thence, in the shape of emancipation, Soul's [chief] End.

Aph. 24.* Bondage [which may be viewed as one of the ends which Soul could arrive at only through the Subtile Body,] is from Misconception.

a. Through the transmigration of the subtile body, from misconception, there is that [less worthy] end of soul, in the shape of bondage, consisting of pleasure and pain: such is the meaning.

b. Liberation and Bondage, [resulting] from knowledge and misconception [respectively], have been mentioned. Of these, in the first place, he explains Liberation [arising] from knowledge:

Aph. 25.* Since this [*viz*., knowledge,] is the precise cause [of liberation], there is neither association [of anything else with it, *e.g.*, good works,] nor alternativeness, [*e.g.*, of good works, in its stead].

a. In respect of there being neither association nor alternativeness, he states an illustration:

Aph. 26.* The emancipation of Soul does not depend on

both [knowledge and works, or the like]; as [any end that one aims at is not obtained] from dreams and from the waking state, [together, or alternatively, which are, severally,] illusory and not illusory.

a. But, even if it be so, [some one may say,] there may be association, or alternativeness, of knowledge of the truth with that knowledge which is termed Worship of [the One, all-constitutive, divine] Soul; since there is no *illusoriness* in *this* object of Worship. To this he replies:

Aph. 27.* Even of that other it is not complete.

a. Even of 'that other,' *i.e.*, of the [just-mentioned] object of worship, the non-illusoriness is not complete; because imaginary things, also, enter into [our conception of, and overlie, and disguise,] the object of worship, the [One, all-constitutive] Soul: such is the meaning.

b. He states in what *part* [of it] is the illusoriness of the [object of] Worship, [just referred to]:

Aph. 28.* Moreover, it is in what is *fancied* that it is thus [illusory].

a. That is to say: 'moreover, it is thus,' *i.e.*, moreover, there is illusoriness, in that portion of the thing meditated which [portion of it] is fancied by the Mind, [while it does not exist in reality]; for, the object of worship having been declared in such texts as, 'All this, indeed, is Brahma,' the illusoriness belongs entirely to that portion [of the impure conception of 'the All' which presents itself, to the undiscriminating, under the aspect] of the world.

b. Then what profit is there in Worship? With reference to this, he declares [as follows]:

Aph. 29.* From the achievement of [the worship termed] meditation there is, to the pure [Soul], all [power]; like Nature.

a. Through the effecting of the worship which is termed meditation, there becomes, to the 'pure,' *i.e.*, the sinless, Soul, all power; as belongs to Nature: such is the meaning. That is to say: as Nature creates, sustains, and destroys, so also the Purity of the understanding of the worshipper, by instigating

Nature, creates, &c. [But this is not Liberation, or Soul's chief end.]

b. It has been settled that Knowledge alone is the means of Liberation. Now he mentions the *means* of Knowledge:

Aph. 30.* Meditation is [the cause of] the removal of Desire.

a. That is to say: Meditation is the cause of the removal of that affection of the mind by objects, which is a hinderer of knowledge.

b. With advertence to the fact that knowledge arises from the effectuation of Meditation, and not from merely commencing upon it, he characterizes the effectuation of Meditation:

Aph. 31.* It [Meditation,] is perfected by the repelling of the modifications [of the Mind, which ought to be abstracted from all thoughts of anything].

a. He mentions also the means of Meditation:

Aph. 32.* This [Meditation,] is perfected by Restraint, Postures, and one's Duties.

a. That is to say: Meditation results from the triad, which shall be mentioned, *viz*., Restraint, &c.

b. By means of a triad of aphorisms he characterizes, in order, Restraint, &c.:

Aph. 33.* Restraint [of the breath] is by means of expulsion and retention.

a. That it is 'of the breath' is gathered from the notoriousness [of its being so].

b. He characterizes Postures, which come next in order:

Aph. 34.* Steady and [promoting]* ease is a [suitable] Posture.

a. That is to say: that is a Posture which, being steady, is a cause of pleasure; such as the crossing of the arms.

b. He characterizes one's Duty:

Aph. 35.* One's Duty is the performance of the actions prescribed for one's religious order.

a. Simple.

Aph. 36.* Through Dispassion and Practice.

a. Simply through mere Practice, in the shape of Meditation, accompanied by Dispassion, Knowledge, with its instrument, Concentration, takes place in the case of those who are most competent [to engage in the matter]· such is the meaning. Thus has liberation through knowledge been expounded.

b. After this, the cause of Bondage, *viz.*, Misconception, declared in [the assertion,] 'Bondage is from Misconception,' [§ 24], is to be expounded. Here he first states the nature of Misconception:

Aph. 37.* The kinds of Misconception are five.

a. That is to say: the subdivisions of Misconception, which is the cause of Bondage, are Ignorance, Egoism, Desire, Aversion, and Fear of Dissolution; the five mentioned in the *Yoga*, [see *Yoga Aphorisms*, Book II., § 3].

b. Having stated the nature of Misconception, he states also the nature of its cause, *viz.*, Disability:

Aph. 38.* But Disability is of twenty-eight sorts.

a. Simple; [as explained in the *Yoga*].

b. In a couple of aphorisms he mentions [those] two. Acquiescence and Perfection, on the prevention of which come two sorts of Disability of the Understanding:

Aph. 39.* Acquiescence is of nine sorts.

a. He will, himself, explain how it is of nine sorts.

Aph. 40.* Perfection is of eight sorts.

a. This, also, he will, himself, explain.

b. Of the aforesaid, *viz.*, Misconception, Disability, Acquiescence, and Perfection, since there may be a desire to know the particulars, there is, in order, a quaternion of aphorisms:

Aph. 41.* The subdivisions [of Misconception] are as [declared] aforetime.

a. The subdivisions of Misconception, which, in a general way, have been stated as five, are to be understood to be particularized 'as aforetime,' *i.e.*, just as they have been declared by preceding teachers: they are not explained here, for fear of prolixity: such is the meaning.

Aph. 42.* So of the other [*viz.*, Disability].

a. That is to say: 'so,' *i.e.*, just as aforetime [§ 41], the divisions 'of the other,' *viz.*, of Disability, also, which are twenty-eight, are to be understood, as regards their particularities.

Aph. 43.* Acquiescence is ninefold, through the divisions of 'the internal and the rest.'

a. This aphorism is explained by a memorial verse [No. 50].

Aph. 44.* Through Reasoning, &c., [which are its subdivisions,] Perfection [is eightfold].

a. That is to say: Perfection is of eight kinds, through its divisions, *viz.*, Reasoning, &c. This aphorism, also, 'has been explained in a memorial verse,' [No. 51].

b. But then, how is it said that Perfection consists only of 'Reasoning, &c.,' seeing that it is determined, in all the Institutes, that the eight Perfections, *viz.*, [the capacity of assuming] atomic bulk, &c., result from recitations, austerity, meditation, &c.? To this he replies:

Aph. 45.* Not from any other [than what we have just stated does real Perfection arise; because what does arise therefrom, *e.g.*, from austerities, is] without abandonment of something else, [*viz.*, Misconception].

a. 'From any other,' *i.e.*, from anything different from the pentad, 'Reasoning, &c.,' *e.g.*, from Austerity, &c., there is no real Perfection. Why? 'Without abandonment of something else;' *i.e.*, because *that* Perfection [which you choose to call such] takes place positively without abandonment of something else, *i.e.*, of Misconception: therefore [*that* Perception], since it is no antagonist to mundane existence, is only a *semblance* of a Perfection, and not a real Perfection: such is the meaning.

b. Now the individuated creation, which was mentioned concisely in the assertion, 'There is distinction of individuals through diversity of desert,' [§ 10], is set forth diffusely:

Aph. 46.* [The creation is that] of which the subdivisions are the demons, &c.

a. Supply, such is that creation, of which 'the subdivisions,' the included divisions, are the demons, &c. This is explained in a memorial verse, [No. 53].

b. He states that the aforesaid subdivided creation, also, is for the sake of Soul:

Aph. 47.* From Brahma down to a post, for its [Soul's,] sake is creation, till there be discrimination [between Soul and Nature].

a. He mentions, further, the division of the subdivided creation, in three aphorisms:

Aph. 48.* Aloft, it [the creation,] abounds in [the quality of] Purity.

a. That is to say: 'aloft,' above the world of mortals, the creation has chiefly [the Quality of] Purity.

Aph. 49.* Beneath, it [the creation,] abounds in Darkness.

a. 'Beneath,' that is to say, under the world of mortals.

Aph. 50.* In the midst, it [the creation,] abounds in Passion.

a. 'In the midst,' that is to say, in the world of mortals.

b. But then, for what reason are there, from one single Nature, creations diverse in having, affluently, purity and the rest? With reference to this, he says:

Aph. 51.* By reason of diversity of desert is Nature's [diverse] behaviour; like a born-slave.

a. Just by reason of diverse desert is the behaviour of Nature, as asserted, in the shape of diversity of operation. An illustration of the diversity is [offered in the example], 'like a born-slave.' That is to say: as, of him who is a slave from the embryo-state upwards, there are, through the aptitude arising from the habit of being a dependant, various sorts of behaviour, *i.e.*, of service, for the sake of his master, so [does Nature serve soul in various ways].

b. But then, if the creation aloft is abundant in Purity [the element of joy], since Soul's object is really thereby effected, what need is there of *Liberation*? To this he replies:

Aph. 52.* Even there there is return [to miserable states of existence]: it is to be shunned, by reason of the successive subjections to birth, [from which the inhabitants of Heaven enjoy no immunity].

a. Moreover:

Aph. 53.* Alike [belongs to all] the sorrow produced by decay and death.

a. Common to all alike, those that are aloft and those beneath, beginning with Brahma and ending with a stock, is the 'sorrow produced by decay and death'; therefore, moreover, it [heaven,] is to be shunned: such is the meaning.

b. What need of more? The end is not effected by absorption into the cause, either; as he tells us:

Aph. 54.* Not by absorption into the cause is there accomplishment of the end; because, as in the case of one who has dived, there is a rising again.

a. In the absence of knowledge of the distinction [between Soul and Nature], when indifference towards Mind, &c., has resulted from worship of Nature, then absorption into Nature takes place; for it is declared: 'Through Dispassion there is absorption into Nature.' Even through this, *i.e.*, the absorption into the cause, the end is not gained; 'because there is a rising again; as in the case of one who has dived.' As a man who has dived under water rises again, exactly so do Souls which have been absorbed into Nature reappear, [at the commencement of a new annus magnus], in the condition of Lords; because it is impossible that one's Faults should be consumed, without a familiarity with the distinction [between Soul and Nature], in consequence of the reappearance of Passion, by reason of the non-destruction of habits, &c.: such is the meaning.

b. But then, the cause is not by anyone caused to act. Being independent, then, why does she [Nature,] make that grief-occasioning resurrection of her own worshipper? To this he replies:

Aph. 55.* Though she be not constrained to act, yet this is fitting; because of her being devoted to another.

a. Though Nature is 'not constrained to act,' not instigated, not subject to the will of another, yet 'this is fitting;' it is proper that he who is absorbed in her should rise again. Why? 'Because of her being devoted to another;' *i.e.*, because she seeks Soul's end. The meaning is that he who is absorbed in her is again raised up, by Nature, for the sake of Soul's end, which consists in knowledge of the distinction [between Nature and Soul]. And Soul's end, and the like, are not *constrainers* of Nature, but occasions for the energizing of her whose very being is to energize; so that there is nothing detracted from her independence.

b. He mentions, further, a proof that Soul rises from absorption into Nature:

Aph. 56.* [He who is absorbed into Nature must rise again;] for he becomes omniscient and omnipotent [in a subsequent creation].

a. For 'he,' *viz.*, he who, in a previous creation, was absorbed into the Cause, in a subsequent creation becomes 'omniscient and omnipotent;' the Lord, the First Spirit.

b. But then, if that be so, it is impossible to deny a *Lord*, [which, nevertheless, the *Sankhyas* seem to do]. To this he replies.

Aph. 57.* The existence of *such* a Lord is a settled point.

a. It is quite agreed, by all, that there is an emergent Lord, he who had been absorbed into Nature; for the ground of dispute [between *Sankhyas* and the rest,] is altogether about an *eternal* Lord: such is the meaning.

b. He expounds diffusely the motive for Nature's creating, which was mentioned only indicatorily in the first aphorism of the Second Book:

Aph. 58.* Nature's creating is for the sake of another, though it be spontaneous;—for she is not the experiencer;—like a cart's carrying saffron [for the sake of its master].

a. But then, it is quite impossible that Nature, being unintelligent, should be, spontaneously, a creator; for we see that a cart, or the like, operates only by reason of the efforts of another. To this he replies:

Aph. 59.* Though she be unintelligent, yet Nature acts; as is the case with milk.

a. That is to say: as milk, without reference to men's efforts, quite of itself changes into the form of curd, so Nature, although she be unintelligent, changes into the form of Mind, &c., even without the efforts of any other.

b. This is not rendered tautological by this aphorism, 'As the cow for the calf,' [Book II., § 37]; because there the question was only of the operation of *instruments*, and because cows *are* intelligent.

c. By means of the exhibition of another illustration, he mentions the cause of the thing asserted as aforesaid:

Aph. 60.* Or as is the case with the acts [or on-goings]—for we see them—of Time, &c.

a. Or as is the case with the acts [or on-goings,] of Time, &c., the spontaneous action of Nature is proved from what is seen. The action of Time, for example, takes place quite spontaneously, in the shape of one season's now departing and another's coming on: let the behaviour of Nature, also, be thus; for the supposition conforms to observed facts: such is the meaning.

b. But, still, a senseless Nature would never energize, or would energize the wrong way; because of there being [in her case,] no such communing as, 'This is my means of producing experience, &c.' To this he replies:

Aph. 61.* From her own nature she acts, not from thought; like a servant.

a. That is to say: as, in the case of an excellent servant, naturally, just from habit, the appointed and necessary service of the master is engaged in, and not with a view to his own enjoyment, just so does Nature energize from habit alone.

Aph. 62.* Or from attraction by Deserts, which have been from eternity.

a. Here the word 'or' is for connecting [this aphorism with the preceding one]. Since Desert has been from eternity, therefore, moreover, through attraction by Deserts, the energizing of Nature is necessary and rightly distributed: such is the meaning.

b. It being thus settled, then, that Nature is creative for the sake of another, he tells us, in the following section, that, on the completion of that other's purpose, Liberation takes place through Nature's quite spontaneously ceasing to act:

Aph. 63.* From discriminative knowledge there is a cessation of Nature's creating; as is the case with a cook, when the cooking has been performed.

a. When Soul's aim has been accomplished, by means of indifference to all else, through discriminative knowledge of Soul, Nature's creating ceases; as, when the cooking is completed, the labour of the cook ceases: such is the meaning.

b. But, at that rate, since Nature's creating ceases through the production of discriminative knowledge in the case of a single Soul, we should find *all* liberated. To this he replies:

Aph. 64.* Another remains like another, through her fault.

a. But 'another,' *i.e.*, one *devoid* of discriminative knowledge, remains 'like another,' *i.e.*, just like one bound by Nature. Why? 'Through her fault,' *i.e.*, through the fault which may be described as her not accomplishing that soul's aim: such is the meaning.

b. He mentions the fruit of Nature's ceasing to act:

Aph. 65.* [The fruit of Nature's ceasing to act], the solitariness of both [Nature and Soul], or [which comes to the same thing,] of either, is liberation.

a. 'Of both,' *i.e.*, of Nature and Soul, the 'solitariness,' *i.e.*, the being alone, the mutual disjunction, in short, this is liberation.

b. But then, how would Nature, having attained indifference, through the mood in the shape of discrimination, on the liberation of a single Soul, again engage in creation, for the sake of another Soul? And you are not to say that this is no objection, because Nature consists of different portions, [it is not *another* Nature, but the same]; because we see, that, even out of the [mortal] constituents of the liberated person, *viz.*, his dust, &c., things are created for the experience of another. To this he replies:

Aph. 66.* Moreover, [when Nature has left off distressing the emancipated,] she does not desist, in regard to her creative influence on another; as is the case with the snake, [which ceases to be a terror,] in respect of him who is aware of the truth in regard to the rope [which another mistakes for a snake].

a. Nature, though, in respect of one Soul, she have desisted, in consequence of discriminative knowledge, does not desist as regards her creative influence on another Soul, but *does* create in respect of *that* one; as the snake [so to speak,] does not produce fear, &c., in the case of him who is aware of the truth in regard to the rope, but *does* produce it, in respect of him who is ignorant [that what he looks upon is a rope, and not a snake]: such is the meaning. And Nature is likened to a snake, because of her disguising Soul, which is likened to a rope. Certain unintelligent persons, calling themselves *Vedantis*, having quite failed to understand that such is the drift of such examples as those of the rope, the snake, &c., suppose that Nature is an absolute nothing, or something merely imaginary. The matters of Scripture and of the legal institutes are to be *elucidated* by means of this [or that] example offered by the *Sankhyas*, who assert the reality of Nature: it is not the case that the matter is simply *established* to be as is the example; [the analogy of which is not to be overstrained, as if the cases were parallel throughout].

Aph. 67.* And from connexion with Desert, which is the cause.

a. 'Desert,' which is the cause of creation, in consequence of the conjunction of this, also, she creates, for the sake of another Soul [than the emancipated one]: such is the meaning.

b. But then, since all Souls are alike indifferent, inasmuch as they do not desire [Nature's interference], what is it that here determines Nature to act only in regard to this one, and to desist in regard to that one? And *Desert* is not the determiner; because here, too, there is nothing to determine of which Soul what is the Desert; [Desert being inferrible only from, and, therefore, not cognizable antecendently to, its *fruits*]. To this he replies:

Aph. 68.* Though there is [on Soul's part, this] indifference, yet want of discrimination is the cause of Nature's service.

a. That is to say: although Souls *are* indifferent, yet Nature, just through [her own] non-discrimination, saying, 'This is my master,' 'This is I myself,' serves Souls, [towards their eventual emancipation], by creation, &c. And so, to what Soul, not having discriminated herself [therefrom], she has the habit of showing herself, in respect just of that one does Nature energize; and this it is that determines her: such is the import.

b. Since it is her *nature* to energize, how can she desist, even when discrimination has taken place? To this he replies:

Aph. 69.* Like a dancer does she, though she had been energizing, desist; because of the end's having been attained.

a. Nature's disposition to energize is only for the sake of Soul, and not universally. Therefore is it fitting that Nature should desist, though she has been energizing, when the end has been attained, in the shape of the effectuation of Soul's aim; as a dancer, who has been performing, with the view of exhibiting a dance to the spectators, desists, on the accomplishment of this: such is the meaning.

b. He states another reason for the cessation:

Aph. 70.* Moreover, when her fault is known, Nature does not approach [Soul]; like a woman of good family.

a. That is to say: Nature, moreover, ashamed at Soul's having seen her fault,—in her transformations, and her taking the shape of pain, &c.,—does not again approach Soul; 'like a woman of good family;' *i.e.*, as a [frail] woman of good family, ashamed at ascertaining that her fault has been seen by her husband, does not approach her husband.

b. But then, if Nature's energizing be for the sake of Soul, Soul must be *altered* by Bondage and Liberation, [and not remain the unalterable entity which you allege it to be]. To this he replies:

Aph. 71.* Bondage and Liberation do not actually belong to Soul, [and would not even appear to do so,] but for non-discrimination.

a. Bondage and Liberation, consisting in the conjunction of Pain, and its disjunction, do not 'actually,' *i.e.*, really, belong to Soul; but, in the way mentioned in the fourth aphorism, they result only from non-discrimination: such is the meaning.

b. But, in reality, Bondage and Liberation, as declared, belong to Nature alone: so he asserts:

Aph. 72.* They really belong to Nature, through consociation; like a beast.

a. Bondage and Liberation, through Pain, really belong to Nature, 'through consociation,' *i.e.*, through her being hampered by the habits, &c., which are the causes of Pain; as a beast, through its being hampered by a rope, experiences Bondage and Liberation: such is the meaning.

b. Here, by what causes is there Bondage? Or by what is there Liberation? To this he replies:

Aph. 73.* In seven ways does Nature bind herself; like the silk-worm: in one way does she liberate herself.

a. By Merit, Dispassion, Supernatural Power, Demerit, Ignorance, Non-dispassion, and Want of Power, *viz.*, by habits, causes of Pain, in the shape of these seven, 'does Nature bind herself' with Pain; 'like the silk-worm;' *i.e.*, as the worm that makes the cocoon binds itself by means of the dwelling which itself constructs. And that same Nature liberates herself from Pain 'in one way,' *i.e.*, by Knowledge alone: such is the meaning.

b. But then, that which you assert, *viz.*, that Bondage and Liberation result from Non-discrimination alone, is improper; because Non-discrimination can neither be quitted nor assumed, and because, in the world, Pain, and its negative, Pleasure, &c., can, themselves, be neither quitted nor assumed: otherwise, [if you still insist on retaining the opinion objected to], there is disparagement of sense-evidence. Having pondered this, he himself [not leaving it to a commentator,] explains what was asserted in the fourth aphorism:

Aph. 74.* Non-discrimination is the cause [not the thing itself]; [so that] there is no disparagement of sense-evidence.

a. What was asserted before was this, that Non-discrimination is only the *occasion* of Bondage and Liberation in souls, and not that Non-discrimination itself is these two; therefore 'there is no disparagement of sense-evidence;' [for, though we see that Pain and Pleasure cannot be directly assumed or quitted, yet we also see that causes of them *can* be assumed or quitted]: such is the meaning.

b. He mentions, among the means conducive to Discrimination, Study, which is the essence of them:

Aph. 75.* Discrimination is perfected through abandonment [of everything], expressed by a 'No, No,' through study of the [twenty-five] Principles.

a. Discrimination is effected through study of the Principles, in the shape of abandoning, by a 'No, No,' in regard to things unintelligent, ending with Nature, the conceit [that Nature, or any of her products, is Soul]. All the others [enumerated in the list of means] are only supplemental to Study: such is the meaning.

b. He states a speciality in regard to the perfecting of Discrimination:

Aph. 76.* Through the difference of those competent [to engage in the matter at all], there is no necessity [that each and every one should at once be successful].

a. Since there is a division, among those competent, into the sluggish, &c., though study be made, there is no certainty that, in this very birth, Discrimination will be accomplished: such is the meaning. Therefore, every one should, by strenuousness in study, acquire for himself the *highest* degree of competency: such is the import.

b. He states that Liberation takes place solely through the effecting of Discrimination, and not otherwise:

Aph. 77.* Since what [Pain] has been repelled returns again, there comes, even from medium [but imperfect,] Discrimination, experience, [which it is desired to get entirely rid of].

a. But sluggish Discrimination [lower even than the middling variety], antecedently to direct intuition, consists only of Hearing, Pondering, and Meditating: such is the division [of Discrimination].

Aph. 78.* And he who, living, is liberated.

a. That is to say: he, also, who, while living, is liberated is just in the condition of medium Discrimination.

b. He adduces evidence for there being some one liberated, though still living:

Aph. 79.* It is proved by the fact of instructed and instructor.

a. That is to say: it is proved that there are such as are liberated during life, by the mention, in the Institutes, on the subject of Discrimination, of the relation of preceptor and pupil; *i.e.*, because it is only one liberated during life that can be an instructor [in this matter].

Aph. 80.* And there is Scripture.

a. There is also Scripture for there being persons liberated during life.

b. But then, merely through hearing, too, one might become [qualified to be] an instructor. To this he replies:

Aph. 81.* [And not through merely hearing is one qualified to become an instructor]: otherwise, there were blind tradition.

a. That is to say: otherwise, since even a person of sluggish Discrimination [but who, yet, had *heard*,] would be an instructor, we should have a blind handing down [of doctrines which would speedily become corrupted or lost].

b. But then, when, through Knowledge, one's works [which are the cause of mundane existence,] have perished how can there [still] be life? To this he replies:

Aph. 82.* Possessed of a body, [the emancipated sage goes on living]; like the whirling of a wheel.

a. Even on the cessation of the action of the potter, the wheel, of itself, revolves for some time, in consequence of the motal inertia resulting from the previous action. So, after knowledge, though actions do not arise, yet, through the [self-continuant] action of antecedent acts, possessing an energizing body, he remains living, yet liberated; [and, if he did not, but if every one who gained true knowledge were, on gaining it, to disappear, true knowledge would cease to be handed down orally; and Kapila, probably, did not contemplate books, or did not think these a secure depository of the doctrine]: such is the meaning.

b. But then, since the continuance of experience, &c., is put an end to by that 'Meditation with distinct recognition of the object,' which [see *Yoga Aphorisms*, Book I., § 17,] is the cause of knowledge, how can one retain a body? To this he replies:

Aph. 83.* This [retention of a body] is occasioned by the least vestige of impression.

a. That is to say: the retention of a body is caused by even the least remains of those impressions of objects which are the causes of having a body.

b. He recapitulates the sense of the declarations of the Institute:

Aph. 84.* That which was to be done has been done, when entire Cessation of Pain has resulted from Discrimination; not otherwise, not otherwise.

a. So much for the Third Book, on Dispassion.

BOOK IV

Now, by means of a collection of narratives, recognized in the Institutes, the means of discriminative knowledge are to be displayed: so, for this purpose the Fourth Book is commenced.

Aph. 1.* As in the case of the king's son, from instruction as to the truth [comes discrimination between soul and Nature].

a. 'Discrimination' is supplied from the concluding aphorism of the preceding section. The meaning is: as, in the case of the king's son, discrimination is produced by instruction as to the truth. The story, here, is as follows: A certain king's son, in consequence of his being born under the [unlucky] star of the tenth portion [of the twenty-seven portions into which the ecliptic is divided], having been expelled from his city, and reared by a certain forester, remains under the idea, that 'I am a forester.' Having learned that he is alive, a certain minister informs him: 'Thou art not a forester; thou art a king's son.' As he, immediately, having abandoned the idean of his being an outcast, betakes himself to his true royal state, saying, 'I am a *king*,' so, too, it [the Soul], in consequence of the instruction of some kind person, to the effect that 'Thou, who didst originate from the First Soul, which manifests itself merely as pure Thought, art [thyself,] a portion thereof,' having abandoned the idea of its being Nature [or of being something material or phenomenal], rests simply upon its own nature, saying, 'Since I am the son of Brahma, I am, myself, Brahma, and not something mundane, different therefrom:' such is the meaning.

b. He exhibits another story, to prove that even women, *shudras*, &c., may gain the [one desirable] end, through a Brahman, by hearing the instructions of a Brahman:

Aph. 2.* As in the case of the goblin, even when the instruction was for the sake of another, [the chance hearer may be benefited].

a. That is to say: though the instruction in regard to the truth was being delivered, by the venerable *Krishna*, for Arjuna's benefit, knowledge of the distinction [between Soul and Nature] was produced in the case of a goblin standing near [and overhearing the discourse]: and so it may happen in the case of others, too.

b. And, if knowledge is not produced from *once* instructing, then a repetition of the instruction is to be made; to which effect he adduces another story:

Aph. 3.* Repetition [is to be made], if not, from once instructing, [the end be gained].

a. That is to say: a repetition of instruction, also, is to be made; because, in the *Chhandogya* [*Upanishad*], and the like; there is mention of AruIi, and others, as having more than once instructed shwetaketu and others.

b. With a view to the removal of desire, he sets forth, with an illustration, the fragility, &c., of Soul's accompaniments:

Aph. 4.* As in the case of father and son; since both are seen; [to one, to die, and the other, to be born].

a. That is to say: Discrimination takes place, through dispassion, in consequence of its being inferred, in respect of one's own self, also, that there is death and birth; since these are seen in the case of father and son. This has been stated as follows: 'The coming into being, and the departure, of Soul [entangled in Nature], may be inferred from [the case of] father and son.'

b. He next explains, by illustrative stories, the subservients to the *perfecting* of knowledge in him in whom knowledge has arisen, and who is devoid of passion:

Aph. 5.* One experiences pleasure or pain [alternatively], from [voluntary] abandonment or [forcible] separation; as in the case of a hawk.

a. That is to say: since people become happy by the abandonment of things, and unhappy by [forcible] separation from them, acceptance of them ought not to be made; 'as in the case of a hawk.' For a hawk, when he has food [before him], if he be driven away by any one, is grieved at being separated from the food; [but] if, of his own accord, he leaves it, then he is free from grief.

Aph. 6.* As in the case of a snake and its skin.

a. That is to say: as a snake readily abandons its old skin, from knowing that it ought to be quitted, just so he who desires liberation should abandon Nature, experienced through a long period, and effete, when he knows that it ought to be quitted. Thus it has been said: 'As a snake... its old skin,' &c.

b. And, when abandoned, he should not again accept Nature and the rest. So, in regard to this, he says:

Aph. 7.* Or as an amputated hand.

a. As no one takes back again an amputated hand, just so this [Nature], when abandoned, he should not readmit: such is the meaning. The word 'Or' is used in the sense of moreover;' [the import of the conjunction being superadditive, not alternative].

Aph. 8.* What is not a means [of liberation is] not to be thought about, [as this conduces only] to bondage; as in the case of Bharata.

a. That which is not an immediate cause of Discrimination, even though it may be a duty, still is 'not to be thought about;' *i.e.*, intention of the mind towards the performance thereof is not to be made; since it tends to Bondage, from its making us forget Discrimination. 'As in the case of Bharata:' that is to say, as was the case with the royal sage Bharata's cherishing Dinanatha's fawn, though [this was] in accordance with duty.

Aph. 9.* From [association with] many there is obstruction to concentration, through passion, &c.; as in the case of a girl's shells.

a. Association is not to be made with many; because, when there is association with many, there is disturbance, through the manifestation of Passion, &c., which destroys concentration; as a jingling is produced by the mutual contact of the shells on a girl's wrist: such is the meaning.

Aph. 10.* Just so, from [the company of] two, also.

a. Just so, even from two there is obstruction to concentration; therefore one ought to abide quite *alone*: such is the meaning.

Aph. 11.* He who is without hope is happy; like Pingala.

a. Having abandoned hope, let a man become possessed of the happiness called contentment; 'like Pingala;' that is to say, as the courtesan called Pingala, desiring a lover, having found no lover, being despondent, became happy, when she had left off hoping.

b. But then, granting that Pain may cease, on the cessation of hope, yet how can there be *happiness*, in the absence of causes thereof? It is replied: That natural happiness, resulting from the predominance of Purity in the mind, which remains obscured by hope, itself resumes its influence, on the departure of hope; as is the case with the coolness of water which [supposed natural coolness] had been hindered [from manifesting itself,] by heat: there is not, in this case, any need of *means*. And it is laid down that precisely this is happiness of Soul.

c. Since it is an obstructer of Concentration, exertion with a view to experience is not to be made, since this will be effected quite otherwise; as he states:

Aph. 12.* [One may be happy,] even without exertion; like a serpent happy in another's house.

a. Supply, 'he may be happy.' The rest is simple. So it has been said: 'The building of a house is, assuredly, painful, and in no way pleasant. A serpent, having entered the dwelling made by another [*e.g.*, a rat], does find confort.'

b.From Institutes, and from preceptors, only the essence is to be accepted; since, otherwise, it may be impossible to concentrate the attention, from there being, by reason of implications, discussions, &c., discrepancies in declared unessential parts, and from the multiplicity of topics. So he says:

Aph. 13.* Though he devote himself to many Institutes and teachers, a taking of the essence [is to be made]; as is the case with the bee.

a. Supply 'is to be made.' The rest is simple. Thus it has been said: 'From small Institutes, and from great, the intelligent man should take, from all quarters, the essence; as the bee does from the flowers.'

b. Be the other means what they may, the direct possession of Discrimination is to be effected only by intentness, through maintaining Meditation; as he tells us:

Aph. 14.* The Meditation is not interrupted of him whose mind is intent on one object; like the maker of arrows.

a. As, in the case of a maker of arrows, with his mind intent solely on the making of an arrow, the exclusion of other thoughts is not interrupted even by a king's passing at his side, so, too, of him whose mind is intent on one point there is in no way an 'interruption of meditation,' *i.e.*, a failure to exclude other thoughts.

Aph. 15.* Through transgression of the enjoined rules there is failure in the aim; as in the world.

a. Whatever rule, for the practisers of Concentration, has been laid down in the Institutes, if it be transgressed, then the end, *viz.*, the effecting of knowledge, is not attained. 'As in the world.' That is to say: just as, in ordinary life, if the enjoined procedures, &c., in regard to a medicine, or the like, be neglected, this or that effect thereof will not be obtained.

b. He states, further, that, if the rules be *forgotten*, the end will not be gained:

Aph. 16.* Moreover, if they be forgotten; as in the case of the female frog.

a. This is plain. And the story of the female frog is this: A certain king, having gone to hunt, saw a beautiful damsel in the forest. And she, being solicited in marriage by the king, made this stipulation: 'When water shall be shown to me by thee, then I must depart.' But, on one occasion, when wearied with sport, she asked the king, 'Where is water?' The king, too, forgetting his agreement, showed her the water. Then she, having become the she-frog *KamarupiGi*, daughter of the king of the frogs, entered the water. And then the king, though he sought her with nets, &c., did not regain her.

b. He mentions a story with reference to the necessity of reflecting on the words of the teacher, as well as hearing them:

Aph. 17.* Not even though instruction be heard is the end gained, without reflexion; as in the case of Virochana.

a. By 'reflexion' is meant such consideration as determines the import of the teacher's words. Without this, though the instruction be heard, knowledge of the truth does not necessarily follow; for it is written, that, though hearing the instruction of Prajapati, Virochana, as between Indra and Virochana, wanted discrimination, from want of reflexion:

Aph. 18.* Of those two, it [reflexion,] was seen in the case of Indra [only].

a. Of those two who are mentioned, [indicated] by the expression 'of those two,' reflexion [was seen, &c.]. And, as between those two, *viz.*, Indra and Virochana, reflexion was seen in the case of Indra: such is the meaning.

b. And he tells us, that, by him who desires to understand thoroughly, attendance on the teacher should be practised for a long time:

Aph. 19.* Having performed reverence, the duties of a student, and attendance, one has success after a long time; as in his case.

a. 'As in his case.' That is to say: as in the case of Indra, so in the case of another, too, only after having practised, under a preceptor, reverence, study of the *Vedas*, service, &c., is there 'success,' *i.e.*, the revelation of truth; not otherwise.

Aph. 20.* There is no determination of the time; as in the Case of Vamadeva.

a. In the arising of knowledge, there is 'no determination of the time,' as, for instance, in its taking place only from causes dependent on the senses. 'As in the Case of Vamadeva.' That is to say: as, in consequence of causes pertaining to a previous life, knowledge arose, in the case of Vamadeva, even when in embryo, so it may in the case of another.

b. But then, since it is written, that the means of knowledge need be nothing other than devotion to those [*viz.*, Brahma, &c.,] who [unlike the Absolute,] have Qualities, knowledge may result from *this*. Why, then, a hard and subtle process of Concentration? To this he replies:

Aph. 21.* Through devotion to something under a superinduced form, [attainment to, or approach towards, knowledge takes place] *by degrees*; as in the case of those who devote themselves to sacrifices.

a. Supply 'there is attainment.' Through devotion to Souls, *e.g.*, Brahma, Vishnu, shiva, under the forms superinduced on them, the effecting of knowledge takes place 'by degrees,' *i.e.*, by the successive attainment of the worlds of Brahma, &c., or else through the purification or the Good principle, &c., but not *directly*; as is the case with sacrificers [whose slaughter of animals, requiring to be expiated, throws them back, so far, in the road to emancipation]: such is the meaning.

b. He tells us, that, moreover, there is no certainty that successive rise to the worlds of Brahma, &c., would effect knowledge:

Aph. 22.* Moreover, after the attainment of what [like the world of Brahma,] is other [than the state of emancipated soul], there is return [to mundane existence]; because it is written [in the 5th *Prapamhaka* of the *Chhandogya Upanishad*]: 'From conjunction with the five fires there is birth,' &c.

a. He exhibits an illustration, to the effect that the effecting of knowledge takes place only in the case of him who is free from passion:

Aph. 23.* By him who is free from passion what is to be left is left, and what is to be taken is taken; as in the case of the swan and the milk.

a. That is to say: only by him who is free from passion is there a quitting 'of what is to be left,' *i.e.*, of Nature, &c., and a taking 'of what is to be taken,' *i.e.*, of Soul; as it is only the swan,—and not the crow, or the like,—that, out of milk and water mingled, by means of leaving the unimportant water, takes the valuable milk, [as the Hindus insist that it does].

b. He tells us that both of these also take place in consequence of association with a perfect man:

Aph. 24.* Or through association with one who has obtained excellence; as in the case thereof.

a. That is to say: moreover, from association with him by whom 'excellence,' *i.e.*, excellence in knowledge, has been obtained, the aforesaid [discrimination] takes place; just as in the case of the swan, [§ 23]; as, in the case of Alarka, Discrimination manifested itself spontaneously, merely through simple association with Dattatreya.

b. He tells us that we ought not to associate with those who are infected with desire:

Aph. 25.* Not of his own accord should he go near one who is infected with desire; like the parrot.

a. Association is not to be made, voluntarily, with a person infected with desire. 'Like the parrot.' That is to say: just as the bird [called a] parrot, by reason of its being exceedingly beautiful, does not [by going near people,] act in a rash manner, through fear of being imprisoned by those who covet it for its beauty.

b. And he states the harm of association with those who labour under desire:

Aph. 26.* [Else he may become] bound, by conjunction with the cords; as in the case of the parrot.

a. And, in the case of associating with those persons, he may become bound, 'by conjunction with the cords,' *i.e.*, by conjunction with their Desire, &c., [the Qualities, punningly compared to cords]; just 'as in the case of the parrot;' that is to say, just as the bird [called a] parrot becomes bound by the cords, *i.e.*, the ropes, of the hunter.

b. He determines, by two [aphorisms]. the means of [effecting] dispassion:

Aph. 27.* Not by enjoyment is desire appeased; as in the case of the saint.

a. That is to say: as, in the case of the saint, Saubhari, desire was not appeased by enjoyment, so, also in the case of others, it is not.

b. But, further:

Aph. 28.* From seeing the fault of both.

a. That is to say: only 'from seeing the fault,' *e.g.*, of being changeable, of consisting of pain, &c., 'of both,' *i.e.*, of Nature and her productions, does the appeasing of desire take place; just as in the case of the saint [§ 27]. For it is written, that Saubhari, just from seeing the evil of society, was afterwards dispassionate.

b. He tells us that incompetency even to accept instruction attaches to him who is infected with the fault of desire, &c:

Aph. 29.* Not in the case of him whose mind is disturbed does the seed of instruction sprout; as in the case of Aja.

a. In him whose mind is disturbed by desire, &c., not even does a sprout spring up from that seed of the tree of knowledge which is in the shape of instruction. 'As in the case of Aja.' That is to say: as not a sprout from the seed of instruction, though delivered to him by Vasishmha, sprang up in the king named Aja, whose mind was disturbed by grief for his wife.

b. What need of more?

Aph. 30.* Not even a mere semblance [of this true knowledge arises in him whose mind is disturbed]; as in the case of a foul mirror.

a. Even superficial knowledge does not arise, from instruction, in one whose mind is disturbed, through the obstruction caused by its wandering away, *e.g.*, to other objects; as an object is not reflected in a foul mirror, through the obstruction caused by the impurities: such is the meaning.

b. Or, if knowledge should spring up in any kind of way, still it may not, he tells us, be in accordance with the instruction:

Aph. 31.* Nor, even though sprung therefrom, is that [knowledge, necessarily,] in accordance therewith; like the lotus.

a. Though sprung 'therefrom,' *i.e.*, from instruction, knowledge is not [necessarily,] in accordance with the instruction, in case this has not been entirely understood. 'Like the lotus.' That is to say: just as the lotus, though the seed be of the best, is not in accordance with the seed, when the mud is faulty. The mind of the student is compared to the mud [in which the lotus-seed was sown].

b. But then, since the Soul's end is, indeed, gained by [the attainment of] supernatural power in the worlds [§ 21. *a*.] of Brahma, &c., to what purpose is the effecting of knowledge, with so much toil, for liberation? To this he replies:

Aph. 32.* Not even on the attainment of glorification has that been done which was to be done; as is the case with the perfection of the objects worshipped, as is the case with the perfection of the objects worshipped.

a. Even though one attain to supernatural power, 'that has not been done which was to be done,' *i.e.*, the end has not been gained; because it is attended by the grief of deficiency and excess. 'As is the case with the perfection of the objects worshipped.' That is to say: as, though the possession of perfection [so called,] belongs to 'the objects worshipped,' *i.e.*, to Brahma, &c., [still] that has not been done which was to be done; since it is written, that even *these*, while in the sleep of Concentration, &c., [still] *practise* Concentration, [from fear of losing what they have attained to]. Just in like manner is the case with him who, by the worship of these, has attained to their supernatural power. Such is the meaning.

b. So much for the Fourth Book, that of Tales, in the Commentary, composed by Vijnana Bhikshu, on Kapila's Declaration of the Samkhya.

BOOK V

a. The tenets of his Institute are completed. Next is begun a Fifth Book, in order to set aside the prima facie notions of others in regard to his Institute. Among those, in the first place he disposes of the objection that the Benediction implied by the expression 'Well,' in the first Aphorism [of Book I.], is purposeless:

Aph. 1.* The [use of a] Benediction [is justified] by the practice of the good, by our seeing its fruit, and by Scripture.

a. The [use of a] Benediction, which we made, is proved to be proper to be made, by these proofs; such is the meaning. The word *iti* is intended to preclude the expectation of any other reasons.

b. He repels those who entertain the prima facie view, that what was asserted in the expression, 'because it is not proved that there *is* a Lord' [see Book I., Aph. 92], is not made out; because [forsooth,] his existence is proved by his being the giver of the fruits of works:

Aph. 2.* Not from its [the world's,] being governed by a

Lord is there the effectuation of fruit: for it is by works [*i.e.*, by merit and demerit,] that this is accomplished.

a. That is to say: it is not proper [to suppose] the effectuation of the change [of the elements] into the shape of the [appropriate] fruit of works, on the ground that the cause is 'governed by a Lord;' because it is possible for the fruit to be effected by the works [*i.e.*, the merit and demerit,] alone, which are indispensable; [and, if we *do* make the additional and cumbrous supposition of a Lord, he cannot reward a man otherwise than according to his works].

b. He declares, further, in [several] aphorisms, that it is not the case that the Lord is the giver of fruit:

Aph. 3.* [If a Lord were governor, then] from intending his own benefit, his government [would be selfish], as is the case [with ordinary governors] in the world.

a. If the Lord were the governor, then his government would be only for his own benefit; as is the case [with ordinary rulers] in the world: such is the meaning.

b. In reply to the doubt, 'grant that the Lord, also, be benefited: what harm?' he says:

Aph. 4.* [He must, then, be] just like a worldly lord, [and] otherwise [than you desire that we should conceive of him].

a. If we agree that the Lord, also, is benefited, he, also, must be something mundane, 'just like a worldly lord;' because, since his desires are [on that supposition,] not [previously] satisfied, he must be liable to grief, &c.: such is the meaning.

b. In reply to the doubt, 'be it even so,' he says:

Aph. 5.* Or [let the name of Lord be] technical.

a. If, whilst there exists also a world, there be a Lord, then let yours, like ours, be merely a technical term for that soul which emerged at the commencement of the creation; since there cannot be an eternal lordship, because of the contradiction between mundaneness and the having an unobstructed will: such is the meaning.

b. He states another objection to the Lord's being the governor:

Aph. 6.* This [position, *viz.*, that there is a Lord,] cannot be established without [assuming that he is affected by] Passion; because that is the determinate cause [of all energizing].

a. That is to say: moreover, it cannot be proved that he is a *governor*, unless there be Passion; because Passion is the determinate cause of activity.

b. But then, be it so, that there is Passion in the Lord, even. To this he replies:

Aph. 7.* Moreover, were that [Pasion] conjoined with him, he could not be eternally free.

a. That is to say: moreover, if it be agreed that there is conjunction [of the Lord] with Passion, he cannot be eternally free; and, therefore, thy tenet [of his eternal freedom] is invalidated.

b. Pray [let us ask], does lordship arise from the immediate union, with Soul, of the wishes, &c., which we hold to be properties of Nature, [not properties of Soul]? Or from an influence by reason of the mere existence of proximity, as in the case of the magnet? Of these he condemns the former alternative:

Aph. 8.* If it were from the conjunction of the properties of Nature, it would turn out that there is association, [which Scripture denies of Soul].

a. From the conjunction, with Soul, of 'the properties of Nature,' *i.e.*, Desire, &c., Soul, also, would turn out [contrary to Scripture,] to be associated with properties.

b. But, in regard to the latter [alternative], he says:

Aph. 9.* If it were from the mere existence [of Nature, not in association, but simply in proximity], then lordship would belong to every one.

a. That is to say: if lordship is by reason of the mere existence of proximity, as in the case of the magnet [which becomes affected by the simple proximity of iron], then it is settled, as we quite intend it should be, that even all men, indifferently, experiencers in this or that [cycle of] creation, [may] have lordship; because it is only by conjunction with all experiencers, that Nature produces Mind, &c. And, therefore, your tenet of there being only one Lord is invalidated.

b. Be it as you allege; yet these are false reasonings; because they contradict the evidence which establishes [the existence of] a Lord. Otherwise, Nature, also, could be disproved by thousands of false reasonings of the like sort. He therefore says:

Aph. 10.* It is not established [that there is an eternal Lord]; because there is no evidence of it.

a. Its establishment, *i.e.*, the establishing that there is an eternal Lord. Of the Lord, in the first place, there is not *sense*-evidence; so that only the evidences of inference and of testimony can be offered; and these are inapplicable: such is the meaning.

b. The inapplicability he sets forth in two aphorisms:

Aph. 11.* There is no inferential proof [of there being a Lord]; because there is [here] no [case of invariable] association [between a sign and that which it might betoken].

a. 'Association,' *i.e.*, invariable concomitancy. 'There is none;' *i.e.*, none exists, [in this case]. And so there is no inferential proof of there being a Lord; because, in such arguments as, 'Mind, or the like, has a maker, because it is a product,' [the fact of] invariable concomitancy is not established, since there is no compulsion [that every product should have had an intelligent maker]. Such is the meaning.

b. Nor, moreover, he tells us, is there [the evidence of] Testimony [to there being a Lord]:

Aph. 12.* Moreover, there is Scripture for [this world's] being the product of Nature, [not of a Lord].

a. Scripture asserts, exclusively, that the world is the product of Nature, not that it has Soul for its cause.

b. He refutes, diffusely, by a cluster [of seven aphorisms], the opinion of an opponent in regard to that which was established in the first Section, *viz.*, 'Bondage does not arise from Ignorance,' [conjoined with Soul].

Aph. 13.* With that which is solitary there cannot be conjunction of the property of Ignorance.

a. Since Soul has no association [with anything whatever], it is plainly impossible for it to be united with the property of Ignorance.

b. But then, [it may be replied,] what is to be asserted is, that the conjunction of Ignorance is simply through force of Ignorance [which is a negation, or nonentity]; and so, since this is no *reality*, there is no *association* occasioned thereby. To this he replies:

Aph. 14.* Since the existence of this [alleged *negative* Ignorance] is established [only] on the ground of its [pretended] conjunction, there is a vicious circle.

a. And, if it is by the conjunction of Ignorance that Ignorance is established, there is 'a vicious circle,' [literally, a resting of each on the other, alternately], a resting a thing on itself; or, in short, a *regressus in infinitum*.

b. In reply to the doubt [suggested by the Naiyayika], 'but then, as in the case of seed and sprout, the *regressus in infinitum* is no objection,' he replies:

Aph. 15.* It is not as in the case of seed and sprout; for Scripture teaches that the world has a beginning.

a. There cannot belong to it such a *regressus in infinitum* as that of seed and sprout; because there is Scripture for the fact that the mundane state of souls, consisting of all undesirable things, *viz*., Ignorance, &c., had a beginning. For we hear, in Scripture, that these cease to exist at the dissolution of all things, in profound sleep, &c. Such is the meaning.

b. But then, [you Vedantis will say], according to us, Ignorance is technically so termed, and is not, *e.g.*, in the shape, specified by the *Yoga*, of supposing what is not soul to be soul; and so, just like your 'Nature,' since this [Ignorance] of ours has an unbroken eternity, though it be lodged in Soul, there is no disparagement of the solitariness thereof: in regard to this doubt, having deliberated on this artificial sense of the word 'Ignorance,' he objects to it:

Aph. 16.* Then Brahma would be found to be excluded [from existence]; because he is something else than knowledge.

a. If the meaning of the word 'Ignorance' (*avidya*) be only 'otherness than knowledge,' then *Brahma*, soul itself, would be found to be excluded, to perish, through his being annihilable by knowledge; since *he* is other than knowledge: such is the meaning. [Further]:

Aph. 17.* Were there not exclusion, then there would be resultlesseness.

a. But, if, the existence of ignorance were really not excluded by knowledge, then there would be resultlessness of knowledge, because of its not debarring Ignorance, [which is the only result competent to knowledge]: such is the meaning.

b. He censures the other alternative, [*viz*., that knowledge *might* exclude Soul]:

Aph. 18.* If it [Ignorance,] meant the being excludible by Knowledge, it would be [predicable], in like manner, of the world, also.

a. If, on the other hand, the being excludible by Knowledge, in the case of the soul, which possesses properties, be, indeed, what is meant by the being Ignorance, in that case 'the world,' the whole mundane system, *viz*., Nature, Mind, &c., would, also, in like manner, be Ignorance. And so, the whole mundane system being merely Ignorance, since the Ignorance would be annihilated by one man's knowledge, the mundane system would become invisible to others, also. Such is the import.

Aph. 19.* If it [Ignorance,] were of *that* nature it would be something that had a commencement.

a. Or suppose it to be the case, that to be Ignorance means simply the being excludible by Knowledge, still such a thing could not have had an *eternal* existence in souls [as held by Vedantis (see § 15, *b*.)], but must have had a commencement. For it is proved, by such recited texts as, 'Consisting of knowledge alone,'12 &c., that, at the time of the universal dissolution, &c., the soul consists of Knowledge alone. Such is the meaning. Therefore, it is settled that there is no other Ignorance, annihilable by Knowledge, than that stated in the *Yoga* system; and this is a property of the understanding only, not a property of the soul.

b. By a cluster of [six] aphorisms, he clears up the prima facie view of an opponent, in regard to that which was stated in the same Book [Book V., § 2], that Nature's energizing is due to Merit:

Aph. 20.* There is no denying Merit; because of the diversity in the operations of Nature.

a. Merit is not to be denied on the ground of its being no object of sense; because it is inferred; since, otherwise, 'the diversity in the operations of Nature' [accommodating one person, and inconveniencing another,] would be unaccounted for: such is the meaning.

b. He states further proof, also:

Aph. 21.* It [the existence of Merit,] is established by Scripture, by tokens, &c.

a. He shows to be a fallacy the argument of the opponent, that Merit exists not, because of there being no sense-evidence of it:

Aph. 22.* There is, here, no necessity; for there is room for other proofs.

a. That is to say: there is no necessity that a thing of which there is no mundane sense-evidence must be non-existent; because things are subject to *other* proofs.

b. He proves that there exists Demerit, as well as Merit:

Aph. 23.* It is thus, moreover, in both cases.

a. That is to say: the proofs apply to Demerit, just as they do to Merit.

Aph. 24.* If the existence [of Merit] be as of course, [because, otherwise, something would be unaccounted for], the same is the case in respect of both.

a. But then, *merit* is proved to exist by a natural consequence in this shape, *viz*., that, otherwise, an injunction would be unaccounted for; but there is none such in respect of *demerit*: so how can Scriptural or logical argument be extended to *demerit*? If any one says this, it is not so; since there is proof, in the shape of natural consequence, 'it is alike, in respect of both,' *i.e.*, of both merit and demerit; because, otherwise, a *prohibitory* injunction, such as, 'He should not approach another's wife,' would be unaccounted for. Such is the meaning.

b. He repels the doubt, that, if Merit, &c., be acknowledged [to exist], then, in consequence of souls' having properties, &c., they must be liable to modification, &c.:

Aph. 25.* It is of the internal organ [not of *soul*] that Merit, &c., are the properties.

a. In the expression '&c.' are included all those that are stated, in the *Vaisheshika* Institute, as peculiar qualities of soul.

b. [To the objection, that the existence of an internal organ, as well as of the Qualities from which such might arise, is debarred by Scripture, be replies]:

Aph. 26.* And of the Qualities, &c., there is not absolute debarment.

a. The Qualities, *viz.*, Purity, &c., and their properties, *viz.*, happiness, &c., and their products, also, *viz.*, Mind, &c., are not denied essentially, but are denied only adjunctively in respect of soul; just as we deny that heat [in red-hot iron,] belongs to the iron.

b. In regard to the doubt, 'Why, again, do we not deny them an essence, as we do to what is meant by the words *sleep*, *wish*, &c.?' he says:

Aph. 27.* By a conjunction of the five members [of an argumentative statement] we discern [that] Happiness [exists].

a. Here, in order to get a particular subject of his assertion, he takes *happiness* alone, one portion of the matter in dispute, as a representative of the entire matter. But the better reading is, 'we discern [that] Happiness, &c., [exist].' The five members of an argumentative statement are the Proposition, Reason, Example, Synthesis [of the two premises], and Conclusion; and, by the 'conjunction,' *i.e.*, the combination, of these, all things, *viz.*, Happiness, &c., are proved to exist. Such is the meaning.

b. And the employment [of the argument] is this:

(1) Pleasure is real;

(2) Because it produces motion in something.

(3) Whatever produces motion in anything is real, as are sentient beings;

(4) And pleasure produces motion in things, in the way of horripilation, &c.:

(5) Therefore, it is real.

c. But then the *Charvaka*, next, doubts whether there be any evidence other than sense-evidence; since [he contends,] there is no truth in the assertion [of an inductive conclusion], that such and such is pervaded by such and such, &c.

Aph. 28.* Not from once apprehending is a connexion established.

a. That is to say: from *once* apprehending concomitance [of a supposed token and the thing betokened], a 'connexion,' *i.e.*, a pervadedness [or invariable attendedness of the token by the betokened,] is not established; and *frequency* [of the same apprehension] follows [the rule of the single apprehension; just as a thousand times nothing amount to nothing]. Therefore [argues the sceptic,] since the apprehending of an invariable attendedness is impossible, nothing can be established by *Inference*. [This] he clears up:

Aph. 29.* Pervadedness is a constant consociation of characters, in the case of both, or of one of them.

a. 'Consociation of characters,' *i.e.*, consociation in the fact of being characters [or properties of something]; in short, concomitancy. And so we mean, that that concomitancy is 'pervadedness,' [furnishing solid ground for inference], which is invariably non-errant, whether in the case of 'both,' the predicate and the reason, or in the case of 'one of them,' the reason only. 'Of both' is mentioned with reference to the case of 'equal pervadedness': [*e.g.*, every equilateral triangle is equiangular, and, conversely, every equiangular triangle is equilateral]. And the invariableness may be apprehended through an appropriate confutation [or *reductio ad absurdum* of the denial of it]; so that there is no impossibility in apprehending 'pervadedness,' [and of inferring on the strength of it]. Such is the import.

b. He declares that Pervadedness is not an additional principle, consisting, *e.g.*, of some such power as is to be mentioned [in § 31]:

Aph. 30.* It [Pervadedness,] is not [as some think (see § 31),] an additional principle [over and above the twenty-five (Book I., § 61)]; for it is unsuitable to postulate entities [*praeter rationem*].

a. 'Pervadedness' is not an entity other than a fixed consociation of characters; because it is unsuitable to suppose, further, some entity as the residence of what constitutes 'pervadedness.' But we consider that what constitutes 'pervadedness' belongs to extant things simply. Such is the meaning.

b. He states the opinion of others:

Aph. 31.* [But certain] teachers say that it [Pervadedness,] is [another principle, in addition to the twenty-five,] resulting from the power of the thing itself.

a. But other teachers assert that 'Pervadedness' is, positively, a separate principle, in the shape of a species of power, generated by the native power of the 'pervaded.' But [they continue,] 'Pervadedness' is not simply a power of the [pervaded] thing itself; else it would exist wherever the thing is, [which 'pervadedness' does *not* do]. For smoke, when it has gone to another place [than the point of its origination], is not attended by fire; and, by going into another place, that power is put an end to. Therefore [contend these teachers,] there is no over-extension in the above-stated definition; for, according to our doctrine, the smoke [which betokens fire] is to be specialized as that which is at the time of origination. Such is the import.

Aph. 32.* Panchashikha says that it ['Pervadedness,'] is the possession of the power of the sustained.

a. That is to say: Panchashikha holds that *pervadingness* is the power which consists in being the *sustainer*, and that

'Pervadedness' is the having the power which consists in being the sustained; for Intellect, and the rest, are treated as being pervaded [or invariably attended,] by Nature, &c.; [and this means that each product, in succession] is *sustained* by what precedes it in the series].

b. But then, why is a 'power of the sustained' postulated? Let 'Pervadedness' be simply an essential power of the *thing* pervaded. To this he [Panchashikha,] replies:

Aph. 33.* The relation is not an essential power; for we should have [in that case,] a tautology.

a. But 'the relation,' *viz.*, 'Pervadedness,' is not an essential power; for we should [thus] have a tautology; because, just as there is no difference between 'water-jar' and 'jar for water,' so, also, there is none in the case of 'Intellect' and 'what is Pervaded' [by Nature, of which Intellect consists]. Such is the meaning.

b. He himself explains the 'Tautology:'

Aph. 34.* Because we should find the distinction unmeaning; [as Intellect does not differ from Nature at all, except as does the sustained from the sustainer].

a. This is almost explained by the preceding aphorism.

b. He [Panchashikha,] mentions another objection:

Aph. 35.* And because it [Pervadedness,] would not be reconcilable in shoots, &c.

a. Because shoots, &c., are invariably attended [at their origination,] by trees, &c. But this cannot be called simply an essential power [in the shoot]; because, since the essential power [that which belongs to the shoot *as being* a shoot,] does not depart, even in the case of an amputated shoot, we should, even then, find it attended [by the tree, which, however, no longer accompanies it]. Such is the sense. But the power [(see § 32), which consists in having the character] of the 'sustained' is destroyed at the time of amputation; so that there is no 'Pervadedness' *then*. Such is the import.

b. But then what? Panchashikha says that 'Pervadedness' is not a result of any *essential* power. Then, since smoke is not *sustained* by fire [see § 32, where he contends that 'sustainedness' is what really expresses *pervasion*], it would turn out that it [*viz.*, smoke,] is not [as token of something that is betokened,] *accompanied* by fire. To this he says:

Aph. 36.* Were it [thus] settled that it is a power of the 'sustained,' then, by the like argument, its dependence on an essential power, [as pretended by the heterodox teachers referred to in § 31, might be proved, also; and thus the argument proves nothing, since it proves too much].

a. That is to say: 'were it settled' that 'a power of the sustained' constitutes the fact of 'Pervadedness,' it would be really settled 'by the like argument.' *i.e.*, by parity of reasoning, that the fact of 'Pervadedness' results from essential power, also, [§ 31. *a*.].

b. It was with a view to substantiate what was stated [in § 27], *viz.*, that the Qualities, and the rest, are established [as realities,] by the employment of the five-membered [form of argumentative exposition], that he has repelled, by an exposition of 'Pervadedness,' the objection to Inference as evidence, [or as a means of attaining right notions].

c. Now, in order to establish the fact that *words*, of which the five-membered [exposition] consists, are generators of knowledge, the objection of others to a *word's* being a means of right knowledge, in the shape of [the objection of] its being inadequate, is disposed of, by means of an exposition of the powers, &c., of words:

Aph. 37.* The connexion between word and meaning is the relation of expressed and expresser.

a. To the 'meaning' belongs the power termed expressibleness; to the 'word,' the power termed expression: simply this is their 'connexion;' their interrelation, as it were. From one's knowing this [connexion between a given word and meaning], the meaning is suggested [or raised in the mind,] by the word. Such is the import.

b. He mentions what things cause one to apprehend the powers [in question]:

Aph. 38.* The connexion [between a word and its sense] is determined by three [means].

a. That is to say: the connexion [just] mentioned [in § 37,] is apprehended by means of these three, *viz.*, information from one competent [to tell us the meaning], the usage of the old man [whose orders to his sons we hear, and then observe what actions ensue, in consequence (see the *Sahitya-darpana*, § 11)], and application to the same thing which has a familiar name, [whence we gather the sense of the less familiar synonym].

Aph. 39.* There is no restriction to what is to be done; because we see it both ways.

a. That is to say: and there is no necessity that this apprehension of the powers [§ 37,] should occur only in the case of 'something [directed] to be done;' because, in [the secular life and dealings of] the world, we see the usage of the old man, &c., [§ 38,] in regard to what is *not* to be done [being something already extant], also, as well as in regard to what is to be done.

Aph. 40.* He who is accomplished in the secular [connexion of words with meanings] can understand the sense of the Veda.

a. Here he entertains a doubt:

Aph. 41.* Not by the three [means mentioned in § 38, objects some one, can the sense of the Veda be gathered]; because the Veda is superhuman, and what it means transcends the senses.

a. Of these he first repels the assertion, that what is meant [by the Veda] is something transcending the senses:

Aph. 42.* Not so [*i.e.*, what is meant by the Veda is not something transcending the senses]; because sacrificings, &c., are, in themselves, what constitutes merit, preeminently.

a. What is asserted [in § 41,] is not the case; since sacrificings, gifts, &c., in the shape, *e.g.*, of the relinquishment of some thing for the sake of the gods, are really, in themselves, 'what constitutes merit,' *i.e.*, what is enjoined by the Veda, 'preeminently,' *i.e.*, because of their having preeminent fruit. And sacrificings, &c., since they are in the shape of wishings, &c., [of which we are perfectly conscious,] are not something transcending intuition. But 'what constitutes merit' [which the objector supposes to transcend intuition,] does not belong to something mysterious that resides in sacrificings &c., whence what is enjoined in the Veda must be beyond intuition. Such is the meaning.

b. He repels also what was asserted [in § 41], *viz.*, that, inasmuch as it [the Veda,] is superhuman, there can be no instruction by any competent person, [in regard to its import]:

Aph. 43.* The natural force [of the terms in the Veda] is ascertained through the conversancy [therewith of those who successively transmit the knowledge].

a. But then, still, how can there be apprehension of the sense of Vaidic terms, in the case of gods, fruits [of actions], &c., which transcend sense? To this he replies:

Aph. 44.* This really takes place; because they [*viz.*, the words,] give rise to knowledge, in the case both of things adapted [to sense] and of things not [so] adapted.

a. He defines the, peculiarities which belong to words, just because this matter is connected with the question of the power of words to cause right knowledge:

Aph. 45.* The Vedas are not from eternity; for there is Scripture for their being a production.

a. Then are the Vedas the work of [the Supreme] Man? To this he replies, 'No':

Aph. 46.* They [the Vedas,] are not the work of [the Supreme] Man; because there is no such thing as the [Supreme] Man, [whom you allude to as being, possibly,] their maker.

a. Supply, 'because we deny that, there is a *Lord*.' [This is] simple.

b. Adverting to the anticipation that there may be some other author, he says:

Aph. 47.* Since the liberated is unsuited [to the work, by his indifference], and the unliberated is so, [by his want of power] neither of these can be author of the Vedas].

a. But then, in that case, since they are not the work of

[the Supreme] Man, it follows that they are eternal. To this he replies:

Aph. 48.* As in the case of sprouts, &c., their eternity does not follow from their not being the work of [any Supreme] Man.

a. [This is] plain.

b. But then, since sprouts, &c., also, just like jars, &c., are productions, we must infer that they are the work of [the Supreme] Man. To this he replies:

Aph. 49.* Were this the case with these, also, [*i.e.*, if it were the case that vegetables were works], we should find a contradiction to experience, &c.

a. It is seen, in the world, as an invariable fact, that whatever is the work of Man is produced by a *body*. This would be debarred, &c., were the case as you contend; [for we see no *embodied* Supreme Man to whose handiwork the sprouts of the earth can be referred]. Such is the meaning.

b. But then, since they were uttered by the Primal Man, the Vedas, moreover, are, really, the work of [the Supreme] Man. To this he replies:

Aph. 50.* That [only] is Man's work, in respect of which, even be it something invisible, an effort of understanding takes place.

a. As in the case of what is visible, so, too, in the case of what is invisible, in respect of what thing there takes place 'an effort of understanding,' *i.e.*, a consciousness that Thought preceded, that thing alone is spoken of as Man's work: such is the meaning. Thus it has been remarked that a thing is not Man's work merely through its having been uttered by Man; for no one speaks of the respiration during profound sleep as being Man's work, [or voluntary act]. But what need to speak of antecedence of Understanding? The Vedas, just like an expiration, proceed, of themselves, from the Self-existent, through the force of fate, wholly unpreceded by thought. Therefore, they are not [a Supreme] Man's work.

b. But then, in that case, since they are not preceded by a correct knowledge of the sense of the sentences, the Vedas, moreover, like the speech of a parrot, can convey no right knowledge. To this he replies:

Aph. 51.* They are, spontaneously, conveyers of right knowledge, from the patentness of their own power [to instruct rightly].

a. That is to say: the authoritativeness of the very whole of the Vedas is established, not by such a thing as its being based on the enouncer's knowledge of the truth, but quite 'spontaneously;' because, as for the Vedas' 'own,' *i.e.*, natural, power of generating right knowledge, *thereof* we perceive the manifestation in the invocations [which produce the result promised], and in the Medical Scripture, [the following of which leads to cures], &c. And so there is the aphorism of the *Nyaya* [Book II., § 681]: 'And [the fact of] its being a cause of right knowledge, like the validity of invocations, and the Medical Scripture,' &c.

b. In regard to the proposition [laid down in § 26, *viz.*], 'And of the [existence of the] Qualities, &c., there is not absolute debarment,' there was duly alleged, and developed [under § 27], one argument, *viz.*, by the establishing the Existence of Happiness, &c. Now he states another argument in respect of that [same proposition]:

Aph. 52.* There is no Cognition of what is no entity, as a man's horn.

a. Be it, moreover, that the existence of pleasure, &c., is proved by the reasoning [under § 27]; it is proved by mere consciousness, also. Of pleasure, &c., were they absolutely *nonentities*, even the *consciousness* could not be accounted for; because there is no cognition of a man's horn, and the like. Such is the meaning.

b. But then, [interposes the *Naiyayika*,] if such be the case, let the Qualities, &c., be quite absolutely *real*; and then, in the expression 'not *absolute* debarment' [in § 26], the word 'absolute' is [superfluous, and, hence,] unmeaning. To this he replies:

Aph. 53.* It is not of the *real* [that there is here cognizance]; because exclusion is seen [of the Qualities].

a. It is not proper [to say], moreover, that the cognizance of the Qualities, &c., is that of the absolutely real; because we see that they are excluded [and not admitted to exist,] at the time of destruction [of the mundane system], &c.

b. But then, even on that showing, let the world be different both from real and from unreal; nevertheless, the demurring to absolute debarment [in § 26,] is untenable. To this he replies:

Aph. 54.* It is not of what cannot be [intelligibly] expressed [that there is cognizance]; because there exists no such thing.

a. And there takes place, moreover, no cognizance of such [a thing] as is not to be expressed as either existing or not existing; 'because there. exists no such thing,' *i.e.*, because nothing is known other than what exists, or what does not exist: such is the meaning. The import is, because it is proper to form suppositions only in accordance with what is seen.

b. But then, on that showing, do you really approve of [the *Nyaya* notion of] 'cognizing otherwise,' [or our fancying that nature to belong to one, which belongs to another]? He replies, 'No':

Aph. 55.* There is no such thing as cognizing otherwise [or cognizing that as belonging to one, which belongs to another]; because your own proposition is self-destructive.

a. This, also, is not proper [to be said], *viz*., that one thing appears under the character of another thing [*e.g.*, a rope, under the character of a serpent, for which it may be mistaken, in the dusk]; 'because your own proposition is self-destructive.' Of another nature [*e.g.*, snakehood], in a different thing [*e.g.*, a rope], *equivalence to a man's horn*, is [what is virtually] expressed by the word 'otherwise' [than the truth; both a man's horn, and the presence of snakehood in a rope mistaken for a snake, being, alike, otherwise than real]; and [yet] its *cognition* [thus] *otherwise* is asserted, [as if *that* could be cognized which is equivalent to what can *not* be cognized]: hence your own proposition is self-destructive. For even those who contend for 'cognizing otherwise' [as one mode of cognition,] declare that the cognition of what *does not exist* is impossible. Such is the meaning.

competent (Adhikari)to study Vedanta and to follow high mental rituals and Yoga processes. Others are not. Some are grown-up children and must be dealt with as such. As all men, and indeed all beings, are, as to their psychical and physical bodies, made of the primordial substance Prakriti-Shakti (Prakrityatmaka), as Prakriti is Herself the three Gunas, Sattva, Rajas and Tamas, and as all things and beings are composed of these three Gunas in varying proportions, it follows that men are divisible into three general classes, namely, those in which the Sattva, Rajas, and Tamas Gunas, predominate respectively. There are, of course, degrees in each of these three classes. Amongst Sattvika men, in whom Sattva predominates, some are more and some less Sattvika than others and so on with the rest.

These three classes of temperament (Bhava)are known in the Shakta Tantras as the Divine (Divyabhava), Heroic (Virabhava)and Animal (Pashubhava)temperaments respectively. Bhava is defined as a property or quality (Dharma)of the Manas or mind (Pranatoshini, 570). The Divyabhava is that in which Sattva-guna predominates only, because it is to be noted that none of the Gunas are, or ever can be, absent. Prakriti cannot be partitioned. Prakriti is the three Gunas. Sattva is essentially the spiritual Guna, for it is that which manifests Spirit or Pure Consciousness (Cit). A Sattvika man is thus a spiritual man. His is a calm, pure, equable, refined, wise, spiritual temperament, free of materiality and of passion, or he possesses these qualities imperfectly, and to the degree that he possesses them he is Sattvik. Pashubhava is, on the other hand, the temperament of the man in whom Tamas guna prevails and produces such dark characteristics as ignorance, error, apathy, sloth and so forth.

He is called a Pashu or animal because Tamas predominates in the merely animal nature as compared with the disposition of spiritually-minded men. He is also Pashu because he is bound by the bonds (Pasha). The term pasha comes from the root Pash to bind. The Kularnava enumerates eight bonds, namely, pity (Daya, of the type which Taoists call "inferior benevolence" as opposed to the divine compassion or Karuna), ignorance and delusion (Moha), fear (Bhaya), shame (Lajja), disgust (Ghrina), family (Kula), habit and observance (Shila), and caste (Varna). Other larger enumerations are given. The Pashu is the man caught by the world, in ignorance and bondage. Bhaskararaya, on the Sutra "have no converse with a Pashu," says that a Pashu is Bahirmukha or outward looking, seeing the outside only of things and not inner realities. The injunction, he says, only applies to converse as regards things spiritual.

The Shaiva Shastra speaks of three classes of Pashu, namely, Sakala bound by the three Pashas, Anu, Bheda, Karma, that is, limited knowledge, the seeing of the one Self as many by the operation of Maya, and action and its product. These are the three impurities (Mala)called Anavamala, Mayamala, and Karmamala. The Sakala Jiva or Pashu is bound by all three, the Pralayakala by the first and last, and the Vijnanakala by the first only. (See as to these the diagram of the 36 Tattvas.)He who is wholly freed of the remaining impurity of Anu is Shiva Himself. Here however Pashu is used in a different sense, that is, as denoting the creature as contrasted with the Lord (Pati). In this sense, Pashu is a name for all men.

In the Shakta use of the term, though all men are certainly Pashu, as compared with the Lord, yet as between themselves one may be Pashu (in the narrower sense above stated)and the other not. Some men are more Pashu than others. It is a mistake to suppose that the Pashu is necessarily a bad man. He may be and often is a good one. He is certainly better than a bad Vira who is really no Vira at all. He is, however, not, according to this Shastra, an enlightened man in the sense that the Vira or Divya is, and he is generally marked by various degrees of ignorance and material-mindedness.

It is the mark of a bad Pashu to be given over to gross acts of sin. Between these two comes the Hero or Vira of whose temperament (Virabhava)so much is heard in the Shakta Shastras. In him there is prevalent the strongly active Rajas Guna. Rajas is always active either to incite Tamas or Sattva. In the former case the result is a Pashu, in the latter case either a Vira or Divya. Where Sattva approaches perfection of development there is the Divyabhava. Sattva is here firmly established in calm and in high degree. But, until such time, and whilst man who has largely liberated himself through knowledge of the influence of Tamas, is active to promote Sattva, he is a Vira. Being heroic, he is permitted to meet his enemy Tamas face to face, counter-attacking where the lower developed man flees away. It has been pointed out by Dr. Garbe (Philosophy of Ancient India, 481), as before him by Baur, that the analogous Gnostic classification of men as material, psychical, and spiritual also corresponds (as does this)to the three Gunas of the Samkhya Darshana.

Even in its limited Shakta sense, there are degrees of Pashu, one man being more so than another. The Pashas are the creations of Maya Shakti. The Devi therefore is pictured as bearing them. But as She is in Her form as Maya and Avidya Shakti the cause of bondage, so as Vidya Shakti She breaks the bonds (Pashupasa-Vimocini)(see v. 78, Lalita-sahasranama), and is thus the Liberatrix of the Pashu from his bondage.

Nitya Tantra says that the Bhava of the Divya is the best, the Vira the next best, and Pashu the lowest. In fact, the state of the last is the starting point in Sadhana, that of the first the goal, and that of the Vira is the stage of one who having ceased to be a Pashu is on the way to the attainment of the goal. From being a Pashu, a man rises in this or some other birth to be a Vira and Divyabhava or Devata-bhava is awakened through Virabhava. The Picchila Tantra says (X, see also Utpatti Tantra, LXIV)that the difference between the Vira and the Divya lies in the Uddhatamanasa, that is, passionateness or activity by which the former is characterized, and which is due to the great effort of Rajas to procure for the Sadhaka a Sattvik state. Just as there are degrees in the Pashu state, so there are classes of Viras, some being higher than others.

The Divya Sadhaka also is of higher or lower kinds. The lowest is only a degree higher than the best type of Vira. The highest completely realize the Deva-nature wherein Sattva exists in a state of lasting stability. Amongst this class are the Tattvajnani and Yogi. The latter are emancipated from all ritual. The lower Divya class may apparently take part in the ritual of the Vira. The object and end of all Sadhana, whether of Pashu or Vira or Divya, is to develop Sattvaguna. The Tantras give descriptions of each of these three classes. The chief general distinction, which is constantly repeated, between the pure Pashu (for there are also Vibhavapashus)and the Vira, is that the former does not, and the latter does, follow the Pancatattva ritual, in the form prescribed for Viracara and described in the next Chapter. Other portions of the description are characteristics of the Tamasik character of the Pashu. So Kubjika Tantra (VII)after describing this class of man to be the lowest, points out various forms of their ignorance.

So it says that he talks ill of other classes of believers. That is, he is sectarian-minded and decries other forms of worship than his own, a characteristic of the Pashu the world over. He distinguishes one Deva from another as if they were really different and not merely the plural manifestations of the One. So, the worshipper of Rama may abuse the worshipper of Krishna, and both decry the worship of Shiva or Devi. As the Veda says, the One is called by various names. Owing to his ignorance "he is always bathing," that is, he is always thinking about external and ceremonial purity. This, though good in its way, is nothing compared with internal purity of mind. He has ignorant or wrong ideas, or want of faith, concerning (Shakta)Tantra Shastra, Sacrifices, Guru, Images, and Mantra, the last of which he thinks to be mere letters only and not Devata (see Pranatoshini, 547, et seq., Picchila, X). He follows the Vaidik rule relating to Maithuna on the fifth day when the wife is Ritusnata (Ritu-kalam vina devi ramanam parivrajayet). Some of the descriptions of the Pashu seem to refer to the lowest class. Generally, however, one may say that from the standpoint of a Viracari, all those who follow Vedacara, Vaishnavacara and Shaivacara are Pashus.

The Kubjika Tantra (VII)gives a description of the Divya. Its eulogies would seem to imply that in all matters which it mentions, the Pashu is lacking. But this, as regards some matters, is Stuti (praise)only. Thus he has a strong faith in Veda, Shastra, Deva and Guru, and ever speaks the truth which, as also other good qualities, must be allowed to the Pashu. He avoids all cruelty and other bad action and regards alike both friend and foe. He avoids the company of the irreligious who decry the Devata. All Devas he regards as beneficial, worshipping all without drawing distinctions. Thus, for instance, whilst an orthodox upcountry Hindu of the Pashu kind who is a worshipper of Rama cannot even bear to hear the name of Krishna, though both Rama and Krishna are each Avatara of the same Vishnu, the Divya would equally reverence both knowing each to be an aspect of the one Great Shakti, Mother of Devas and Men.

This is one of the first qualities of the high Shakta worshipper. As a worshipper of Shakti he bows down at the feet of women regarding them as his Guru (Strinam padatalam drishtva guruvad bhava pet sada). He offers everything to the supreme Devi regarding the whole universe as pervaded by Stri (Shakti, not "woman")and as Devata. Shiva is (he knows)in all men. The whole universe (Brahmanda)is pervaded by Shiva Shakti.

The description cited also deals with his ritual, saying that he does daily ablutions, Sandhya, wears clean cloth, the Tripundra mark in ashes or red sandal, and ornaments of Rudraksha beads. He does Japa (recitation of Mantras external and mental)and worship (Arcana). He worships the Pitris and Devas and performs all the daily rites. He gives daily charity. He meditates upon his Guru daily, and does worship thrice daily and, as a Bhairava, worships Parameshvari with Divyabhava. He worships Devi at night

(Vaidik worship being by day), and after food (ordinary Vaidik worship being done before taking food). He makes obeisance to the Kaula Shakti (Kulastri)versed in Tantra and Mantra, whoever She be and whether youthful or old. He bows to the Kula-trees (Kulavriksha). He ever strives for the attainment and maintenance of Devatabhava and is himself of the nature of a Devata.

Portions of this description appear to refer to the ritual and not Avadhuta Divya, and to this extent applicable to the high Vira also. The Mahanirvana (I. 56) describes the Divya as all but a Deva, ever pure of heart, to whom all opposites are alike (Dvandvatita)such as pain and pleasure, heat and cold, who is free from attachment to worldly things, the same to all creatures and forgiving. The text I have published, therefore, says that there is no Divya-bhava in the Kaliyuga nor Pashubhava; for the Pashu (or his wife)must, with his own hand, collect leaves, flowers and fruit, and cook his food, which regulations and others are impossible or difficult in the Kali age.

As a follower of Smriti, he should not "see the face of a Shudra at worship, or even think of woman" (referring to the Pancatattva ritual). The Shyamarcana (cited in Haratattvadidhiti, 348)speaks to the same effect. On the other hand, there is authority for the proposition that in the Kaliyuga there is only Pashubhava. Thus, the Pranatoshini (510-517)cites a passage purporting to come from the Mahanirvana which is in direct opposition to the above:

Divpa-vira-mayo bhavah kalau nasti kadacana

Kevalampashu-bhavena mantra-siddhir bhaven nrinam.

(In the Kali age there is no Divya or Virabhava. It is only by the Pashu-bhava that men may attain Mantra-siddhi.)

I have discussed this latter question in greater detail in the introduction to the sixth volume of the series of "Tantrik Texts".

Dealing with the former passage from the Mahanirvana, the Commentator explains it as meaning "that the conditions and characters of the Kaliyuga are not such as to be productive of Pashubhava, or to allow of its Acara (in the sense of the strict Vaidik ritual). No one, he says, can now-a-days fully perform the Vedacara, Vaishnavacara, and Shaiva-cara rites without which the Vaidik and Pauranic Yajna and Mantra are fruitless. No one now goes through the Brahmacarya Ashrama or adopts, after the fiftieth year, Vanaprastha.

Those whom the Vaidik rites do not control cannot expect the fruit of their observances. On the contrary, men have taken to drink, associate with the low and are fallen, as are also those who associate with them.. There can, therefore, be no pure Pashu. (That is apparently whilst there may be a natural Pashu disposition the Vaidik rites appropriate to this bhava cannot be carried out.)Under these circumstances, the duties prescribed by the Vedas which are appropriate for the Pashu being incapable of performance, Shiva, for the liberation of men of the Kali age, has proclaimed the Agama. Now there is no other way."

We are, perhaps, therefore, correct in saying that it comes to this: In a bad age, such as the Kali, Divya men are (to say the least)very scarce, though common-sense and experience must, I suppose, allow for exceptions. Whilst the Pashu natural disposition exists, the Vaidik ritual which he should follow cannot be done. It is in fact largely obsolete. The Vaidik Pashu or man who followed the Vaidik rituals in their entirety is non-existent. He must follow the Agamic rituals which, as a fact, the bulk of men do. The Agama must now govern the Pashu, Vira and would-be Divya alike. As I have frequently explained, there are various communities of the followers of Tantra of Agama according to the several divisions of the worshippers of the five Devatas (Pancopasaka). Of the five classes, the most important are Vaishnava, Shaiva and Shakta. I do not, however, hesitate to repeat a statement of a fact of which those who speak of "The Tantra" ignore.

The main elements of Sadhana are common to all such communities following the Agamas; such as Puja (inner and outer), Pratima or other emblems (Linga, Shalagrama), Upacara, Sandhya, Yajna, Vrata, Tapas, Mandala, Yantra, Mantra, Japa, Purashcarana, Nyasa, Bhutasuddhi, Mudra, Dhyana, Samskara and so forth. Even the Vamacara ritual which some wrongly think to be peculiar to the Shaktas, is or was followed (I am told)by members of other Sampradayas including Jainas and Bauddhas. Both, in so far as they follow this ritual, are reckoned amongst Kaulas though, as being non-Vaidik, of a lower class.

A main point to be here remembered, and one which establishes both the historical and practical importance of the Agamas is this: That whilst some Vaidik rites still exist, the bulk of the ritual of to-day is Agamic, that is, what is popularly called Tantrik. The Puranas are replete with Tantrik rituals.

Notwithstanding a general community of ritual forms, there are some variances which are due to two causes: firstly, to difference in the Devata worship, and secondly, to difference of philosophical basis according as it is Advaita, Vishishtadvaita, or Dvaita. The presentment of fundamental ideas is sometimes in different terms. Thus the Vaishnava Pancaratra Agama describes the creative process in terms of the Vyuhas, and the Shaiva-Shakta Agamas explain it as the Abhasa of the thirty-six Tattvas. I here deal with only one form, namely, Shakta Sadhana in which the Ishtadevata is Shakti in Her many forms.

I will here shortly describe some of the ritual forms above-mentioned, premising that so cursory an account does not do justice to the beauty and profundity of many of them.

There are four different forms of worship corresponding to four different states and dispositions (Bhava)of the Sadhaka himself. The realization that the Supreme Spirit (Paramatma)and the individual spirit (Jivatma)are one, that everything is Brahman, and that nothing but the Brahman has lasting being is the highest state or Brahma-bhava. Constant meditation with Yoga-processes upon the Devata in the heart is the lower form (Dhyanabhava). Lower still is that Bhava of which Japa (recitations of Mantra)and Hymns of praise (Stava)are the expression; and lowest of all is external worship (Bahyapuja).

Pujabhava is that which arises out of the dualistic notions of worshipper and worshipped, the servant and the Lord, a dualism which necessarily exists in greater or less, degree until Monistic experience (Advaita-bhava)is attained. He who realizes the Advaita-tattva knows that all is Brahman. For him there is neither worshipper nor worshipped, neither Yoga, nor Puja nor Dharana, Dhyana, Stava, Japa, Vrata or other ritual or process of Sadhana. For, he is Siddha in its fullest sense, that is, he has attained Siddhi which is the aim of Sadhana. As the Mahanirvana says, "for him who has faith in and knowledge of the root, of what use are the branches and leaves'?" Brahmanism thus sagely resolves the Western dispute as to the necessity or advisability of ritual. It affirms it for those who have not attained the end of all ritual. It lessens and refines ritual as spiritual progress is made upwards; it dispenses with it altogether when there is no longer need for it. But, until a man is a real "Knower", some Sadhana is necessary if he would become one. The nature of Sadhana, again, differs according to the temperaments (Bhava)above described, and also with reference to the capacities and spiritual advancement of each in his own Bhava. What may be suitable for the unlettered peasant may not be so for those more intellectually and spiritually advanced. It is, however, a fine general principle of Tantrik worship that capacity, and not social distinction such as caste, determines competency for any particular worship. This is not so as regards the Vaidik ritual proper. One might have supposed that credit would have been given to the Tantra Shastra for this. But credit is given for nothing. Those who dilate on Vaidik exclusiveness have nothing to say as regards the absence of it in the Agama.

The Shudra is precluded from the performance of Vaidik rites, the reading of the Vedas, and the recital of Vaidik Mantras. His worship is practically limited to that of his Ishtadevata, the Vana-lingapuja with Tantrik and Pauranik mantra and such Vrata as consist in penance and charity. In other cases, the Vrata is performed through a Brahmana. The Tantra Shastra makes no caste distinction as regards worship, in the sense that though it may not challenge the exclusive right of the twice-born to Vaidik rites, it provides other and similar rites for the Shudra. Thus there is both a Vaidik and Tantrik Gayatri and Sandhya, and there are rites available for worshippers of all castes.

All may read the Tantras which contain their form of worship, and carry them out and recite the Tantrik Mantras. All castes, even the lowest Candala may, if otherwise fit, receive the Tantrik initiation and be a member of a Cakra or circle of worship. In the Cakra all the members partake of food and drink together, and are then deemed to be greater than Brahmanas, though upon the break-up of the Cakra the ordinary caste and social relations are re-established. It is necessary to distinguish between social differences and competency (Adhikara)for worship. Adhikara, so fundamental a principle of Brahmanism, means that all are not equally entitled to the same teaching and ritual. They are entitled to that of which they are capable, irrespective (according to the Agama)of such social distinctions as caste. All are competent for Tantrik worship, for, in the words of the Gautamiya which is a Vaishnava Tantra (Chap. I)the Tantra Shastra is for all castes and all women.

Sarva-varnadhikarash ca narinam yog ya eva ca.

Though according to Vaidik usage, the wife was co-operator (Sahadharmini)in the household rites, now-a-days, so far as I can gather, they are not accounted much in such matters, though it is said that the wife may, with the consent of her husband, fast, take vows, perform Homa, Vrata and the like. According to the Tantra Shastra, a woman may not only receive Mantra, but may, as Guru, initiate and give it (see Rudrayamala II, ii, and XV). She is worshipped both as wife of Guru and as Guru herself (see ib., I. i. Matrikabheda Tantra (c. vii), Annadakalpa Tantra cited in Pranatosini, p. 68, and as regards the former Yogini Tantra chap. i. Gurupatni Maheshani gurur eva). The Devi is Herself the Guru of all Shastras and woman, as indeed all females Her embodiments, are in a peculiar sense, Her representatives. For this reason all women are worshipful, and no harm should be ever done them, nor should any female animal be sacrificed.

Puja is the common term for ritual worship, of which there are numerous synonyms in the Sanskrit language such as Arcana, Vandana, Saparyya, Arhana, Namasya, Arca, Bhajana, though some of these stress certain aspects of it. Puja as also Vrata which are Kamya, that is, done to gain a particular end, are preceded by the Sankalpa, that is, a statement of the resolve to worship, as also of the particular object (if any)with which it is done. It runs in the form, "I—of—Gotra and so forth identifying the individual)am about to perform this Puja (or Vrata)with the object—". Thereby the attention and will of the Sadhaka are focused and braced up for the matter in hand. Here, as elsewhere, the ritual which follows is designed both by its complexity and variety (which prevents the tiring of the mind)to keep the attention always fixed, to prevent it from straying and to emphasize both attention and will by continued acts and mental workings.

The object of the worship is the Ishtadevata, that is, the particular form of the Deity whom the Sadhaka worships, such as Devi in the case of a Shakta, Shiva in the case of the Shaiva (in eight forms in the case of Ashtamurti-puja as to which see Todala Tantra, chap. V) and Vishnu as such or in His forms as Rama and Krishna in the case of the Vaishnava Sadhaka.

An object is used in the outer Puja (Bahyapuja) such as an image (Pratima), a picture and emblem such as a jar (Kalasa), Shalagrama (in the case of Vishnu worship), Linga and Yoni or Gauripatta (in the case of the worship) of Shiva (with Devi), or a geometrical design called Yantra. In the case of outer worship the first is the lowest form and the last the highest. It is not all who are capable of worshipping with a Yantra. It is obvious that simpler minds must be satisfied with images which delineate the form of the Devata completely and in material form. The advanced contemplate Devata in the lines and curves of a Yantra.

In external worship, the Sadhaka should first worship inwardly the mental image of the Devata which the outer objects assist to produce, and then by the life-giving (Prana-Pratishtha)ceremony he should infuse the image with life by the communication to it of the light, consciousness, and energy (Tejas)of the Brahman within him to the image without, from which there then bursts the luster of Her whose substance is Consciousness Itself (Caitanyamayi). In every place She exists as Shakti, whether in stone or metal as elsewhere, but in matter is veiled and seemingly inert. Caitanya (Consciousness)is aroused by the worshipper through the Pranapratishtha Mantra. An object exists for a Sadhaka only in so far as his mind perceives it. For and in him its essence as Consciousness is realized.

This is a fitting place to say a word on the subject of the alleged "Idolatry" of the Hindus. We are all aware that a similar charge has been made against Christians of the Catholic Church, and those who are conversant with this controversy will be better equipped both with knowledge and caution against the making of general and indiscriminate charges.

It may be well doubted whether the world contains an idolater in the sense in which that term is used by persons who speak of "the heathen worship of sticks and stones". According to the traveler A. B. Ellis ("The Tshi speaking peoples of the Gold Coast of West Africa"), even "negroes of the Gold Coast are always conscious that their offerings and worship are not paid to the inanimate object itself but to the indwelling God, and every native with whom I have conversed on the subject has laughed at the possibility of its being supposed that he would worship or offer sacrifice to some such object as a stone". Nevertheless a missionary or some traveler might tell him that he did. An absurd attitude on the part of the superior Western is that in which the latter not merely tells the coloured races what they should believe, but what notwithstanding denial, they in fact believe and ought to hold according to the tenets of the latter's religion.

The charge of idolatry is kept up, notwithstanding the explanations given of their beliefs by those against whom it is made. In fact, the conviction that Eastern races are inferior is responsible for this. If we disregard such beliefs, then, anything may be idolatrous. Thus; to those who disbelieve in the "Real Presence," the Catholic worshipper of the Host is an idolater worshipping the material substance, bread. But, to the worshipper who believes that it is the Body of the Lord under the form of bread, such worship can never be idolatrous. Similarly as regards the Hindu worship of images. They are not to be held to worship clay or stone because others disbelieve in the efficacy of the Prana-Pratishtha ceremony. When impartially considered, there is nothing necessarily superstitious or ignorant in this rite.

Nor is this the case with the doctrine of the Real Presence which is interpreted in various ways. Whether either rite has the alleged effect attributed to it is another question. All matter is, according to Shakta doctrine, a manifestation of Shakti, that is, the Mother Herself in material guise. She is present in and as everything which exists. The ordinary man does not so view things. He sees merely gross unconscious matter. If, with such an outlook, he were fool enough to worship what was inferior to himself, he would be an idolater. But the very act of worship implies that the object is superior and conscious. To the truly enlightened Shakta everything is an object of worship, for all is a manifestation of God who is therein worshipped. But that way of looking at things must be attained. The untutored mind must be aided to see that this is so. This is effected by the Pranapratishtha rite by which "life is established" in the image of gross matter. The Hindu then believes that the Pratima or image is a representation and the dwelling place of Deity. What difference, it may be asked, does this really make? How can a man's belief alter the objective fact?

The answer is, it does not. God is not manifested by the image merely because the worshipper believes Him to be there. He is there in fact already. All that the Pranapratishtha rite does is, to enliven the consciousness of the worshipper into a realization of His presence. And if He be both in fact, and to the belief of the worshipper, present, then the Image is a proper object of worship. It is the subjective state of the worshipper's mind which determines whether an act is idolatrous or not. The Prana-Pratishtha rite is thus a mode by which the Sadhaka is given a true object of worship and is enabled to affirm a belief in the divine omnipresence with respect to that particular object of his devotion. The ordinary notion that it is mere matter is cast aside, and the divine notion that Divinity is manifested in all that is, is held and affirmed. "Why not then" (some missionary has said)"worship my boot?"

There are contemptible people who do so in the European sense of that phrase. But, nevertheless, there is no reason, according to Shakta teaching, why even his boot should not be worshipped by one who regards it and all else as a manifestation of the One who is in every object which constitutes the Many. Thus this Monistic belief is affirmed in the worship by some Shaktas of that which to the gross and ordinary mind is merely an object of lust. To such minds, this is a revolting and obscene worship. To those for whom such object of worship is obscene, such worship is and must be obscene. But what of the mind which is so purified that it sees the Divine presence in that which, to the mass of men, is an incitement to and object of lust? A man who, without desire, can truly so worship must be a very high Sadhaka indeed. The Shakta Tantra affirms the Greek saying that to the pure all things are pure. In this belief and with, as the as the Jnanarnava Tantra says, the object of teaching men that this is so, we find the ritual use of substances ordinarily accounted impure. The real objection to the general adoption or even knowledge of such rites lies, from the Monistic standpoint, in the fact that the vast bulk of humanity are either of impure or weak mind, and that the worship of an object which is capable of exciting lust will produce it, not to mention the hypocrites who, under cover of such a worship, would seek to gratify their desires.

In the Paradise Legend, just as amongst some primitive tribes, man and woman go naked. It was and is after they have fallen that nakedness is observed by minds no longer innocent. Rightly, therefore, from their standpoint, the bulk of men condemn such worship. Because, whatever may be its theoretical justification under conditions which rarely occur, pragmatically and for the bulk of men they are full of danger. Those who go to meet temptation should remember the risk. I have read that it is recorded of Robert d'Arbrissel, the saintly founder of the community of Fonte d'Evrault that he was wont on occasions to sleep with his nuns, to mortify his flesh and as a mode of strengthening his will against its demands. He did not touch them, but his exceptional success in preserving his chastity would be no ground for the ordinary man undertaking so dangerous an experiment.

In short, in order to be completely just, we must, in individual cases, consider intention and good faith. But, practically and for the mass, the counsel and duty to avoid the occasion of sin is, according to Shastrik principles themselves, enjoined. As a matter of fact, such worship has been confined to so limited a class that it would not have been necessary to deal with the subject were it not connected with Shakta worship, the matter in hand. To revert again to the "missionary's boot": whilst all things may be the object of worship, choice is naturally made of those objects which, by reason of their effect on the mind, are more fitted for it. An image or one of the usual emblems is more likely to raise in the mind of the worshipper the thought of a Devata than a boot, and therefore, even apart from scriptural authority, it would not be chosen. But, it has been again objected, if the Brahman is in and appears equally in all things, how do we find some affirming that one image is more worthy of worship than another. Similarly, in Catholic countries, we find worshippers who prefer certain churches, shrines, places of pilgrimage and representations of Christ, His Mother and the Saints.

Such preferences are not statements of absolute worth but of personal inclinations in the worshipper due to his belief in their special efficacy for him. Psychologically all this means that a particular mind finds that it works best in the direction desired by means of particular instruments. The image of Kali provokes in general only disgust in an European mind. But to the race-consciousness which has evolved that image of Deity, it is the cause and object of fervent devotion. In every case, those means must be sought and applied which will produce a practical and good result for the individual consciousness in question.

It must be admitted, however, that image worship like everything else is capable of abuse; that is a wrong and (for want of a better term)an idolatrous tendency may manifest. This is due to ignorance. Thus the aunt of a Catholic schoolboy friend of mine had a statue of St. Anthony of Padua. If the saint did not answer her prayers, she used to give the image a beating, and then shut it up in a cupboard with its "face to the wall" by way of punishment. I could cite numbers of instances of this ignorant state of mind taken from the past and present history of Europe. It is quite erroneous to suppose that such absurdities are confined to India, Africa or other coloured countries. Nevertheless, we must, in each case, distinguish between the true scriptural teaching and the acts and notions of which they are an abuse.

The materials used or things done in Puja are called Upacara. The common number of these is sixteen, but there are more and less (see Principles of Tantra, Part ii). The sixteen which include some of the lesser number and are included in

Scripture, of only one Subtile Body [made up of the organs], transmigrating generally throngh the different localities. Such is the meaning.

b. But then, in that case, how is the Scripture relating to the materiality [of the organs] to be accounted for? To this he replies:

Aph. 110.* The mention thereof [*viz.*, of materiality, as if it belonged to the organs,] is because there is [intended to be made, thereby, a more emphatic] mention of the concomitant cause.

a. There is designation as the material cause, in the case even where the cause is [but] concomitant, with a view to indicating its *importance*; just as fire is [spoken of as a rising] from fuel, [which fuel is a necessary concomitant of, though not really the substance of, the fire]. Hence are they [the organs,] spoken of as being formed of the Elements. Such is the meaning. For, only in reliance on the support of Light, or other Element, do the Organs, *viz.*, the Sight, &c., [formed] from the accompanying Self-consciousness, come to exist; as fire, in reliance on the support of earthly fuel, results from the attendant Light, [or Heat, which cannot manifest itself alone].

b. As the subject presents itself, he determines the variety that belongs to Gross Body:

Aph. 111.* The heat-born, egg-born, womb-born, vegetable, thought-born, and spell-born; such is not an exhaustive division [of Gross Body, though a rough and customary one].

a. It was stated, before, that Body has only one Element as its material. In this same connexion, he observes discriminatively, as follows:

Aph. 112.* In all [Bodies] Earth is the material: in consideration [however,] of some speciality, there is designation as this [or that other element than earth, as entering into the constitution of some given body], as in the preceding case [treated under § 110].

a. In all Bodies the material is Earth only. 'In consideration of some speciality;' *i.e.*, in consequence of intensity through excess, &c., in the case of Body, as before [in the case of the Organs], there is, however, designation as consisting of Elements, five, or four, &c., on the ground only of there being a *support*, as in the case of the materiality of the Organs. Such is the meaning.

b. But then, since the vital air is the principal thing in the Body, let the vital air itself be the originant of the Body. To this he replies:

Aph. 113.* The vital air is not [on the allegation that it is the principal thing in the Body, to be considered] the originant of the Body; because it [the vital air, or spirit,] subsists through the power of the organs.

a. The vital air, consisting in the function of the organs, does not subsist in the absence of the organs. Therefore, since, in a *dead* Body, in consequence of the absence of the organs, there is the absence of the vital air, the vital air is not the originant of the Body.

b. But then, in that case, since the vital air is not the cause of the Body, the Body might come into existence even without the vital air. To this he replies:

Aph. 114.* The site of experience [*viz*., the Body,] is contructed [only] through the superintendence of the experiencer [Soul]: otherwise, we should find putrefaction.

a. 'Through the superintendence,' *i.e.*, only through the operation, 'of the experiencer,' *i.e.*, Soul [literally, that which has the vital airs], is 'the construction of the site of experience,' *i.e.*, the Body; because, 'otherwise,' *i.e.*, if the operation of the vital airs were absent, we should find putrefaction in the semen and blood, just as in a dead body. Such is the meaning. And thus, by the several operations of circulating the juices, &c., the vital air is a *concomitant* cause of the Body, through the sustaining of it: such is the import.

b. But then [it may be said], it is only the vital air, itself, that can be the superintender; because it is this which operates, not the Soul, since *it* is motoinless, and since there is no use in the superintendence of what does not operate. To this he replies:

Aph. 115.* Through a servant, not directly, is superintendence [exercised] by the master.

a. In the construction of the Body, 'superintendence,' in the shape of energizing, is not 'directly,' *i.e.*, immediately, [exercised] 'by the master,' *i.e.*, by Soul, but 'through its servant,' in the shape of the vital airs; as in the case of a king's building a city: such is the meaning.

b. It was stated before [Book II., § 1,] that Nature's [agency] is 'for the emancipation of what is [really, though not apparently,] emancipated.' In reference to the objection of opponents in regard to this, *viz.*, 'How can the soul be eternally free, when we see it bound?' with a view to demonstrating its eternal freedom, he says:

Aph. 116.* In Concentration, profound sleep, and emancipation, it [Soul,] consists of Brahma.

a. Then what is the difference of emancipation from profound sleep and concentration? To this he replies:

Aph. 117.* In the case of the two, it is with a seed; in the case of the other, this is wanting.

a. 'In the case of the two,' *viz.*, concentration and profound sleep, the identity with Brahma is 'with a seed,' *i.e.*, associated with some cause of Bondage, [or reappearance in the mundane state]; 'in the case of the other,' *i.e.*, in emancipation, this cause is absent: this is the distinction. Such is the meaning.

b. But then, Concentration and profound sleep are *evident*: but what evidence is there of *Emancipation*? This objection of the atheist he repels:

Aph. 118.* But there are not the two [only]; because the triad, also [Emancipation inclusive], is evident; as are the two.

a. The meaning is, that, since Emancipation, also, is 'evident,' *i.e.*, is inferrible, through the example of Concentration and profound sleep, there are not the two, *viz.*, profound sleep and Concentration, only; but Emancipation, also, really is. And the argument is thus. The quitting of that identity with Brahma which [identity] exists during profound sleep, &c., takes place only through a fault, *viz.*, Desire, or the like, lodged in the mind; and, if this fault be annihilated by knowledge, then there results a *permanent* condition, quite similar to profound sleep, &c.; and it is precisely *this* that is Emancipation.

b. But then [suggests some one, with reference to § 117], granting, that, even notwithstanding the existence of the 'seed' [or source of return to the mundane state,] called Memory, a mental modification after the form of any object does not arise during *concentration*, inasmuch as Memory is [then] dulled [or deadened] by apathy, &c., yet, in the case of a person in *profound sleep*, since Memory prevails, there will really be cognition of objects; consequently, it is not proper to say that there is identity with Brahma during profound sleep. To this he replies:

Aph. 119.* There is not the revelation, by memory, of an object likewise during the conjunction of a [more potent] fault [such as sleep]: the secondary cause does not debar the principal.

a. As in the case of apathy, so, also when there is the conjunction of the fault of sleep, Memory does not reveal its own objects, does not remind us of its objects; for the 'secondary,' the subordinate, Memory, cannot defeat the more potent fault of Sleep: such is the meaning. For the really more potent fault makes the memory powerless, incompetent to produce its effects; [and so there is nothing, in this, to prevent identification of Soul with Brahma, during profound sleep, any more than during apathetic Concentration]: such is the import.

b. It was stated, in the Third Book [§ 83], that the retention of a Body by him who is emancipated while still living, is 'in consequenue of a mere vestige of impression.' To this it is objected as follows. Experience is observed, in the case of the [alleged person] emancipated during life, just as in the case of the like of us, [and this experience continuous,] even though it may be constantly in respect of a single object: now, this is unaccountable [on the hypothesis of his really *being* emancipated]; because the antecedent *impression* is annihilated, exactly on its having produced the first [instant of] experience, and because no subsequent impression arises, inasmuch as *knowledge* debars it; just as is the case with Merit. To this he replies:

Aph. 120.* A single impression [suffices to generate, and] lasts out the experience: but there are not different impressions, one to each [instant of] experience; else, we should have a postulation of many, [where a single one may suffice].

a. In like manner, in the case of the whirling of the potter's wheel, the self-continuant principle, called motal inertia, is to be regarded as only one, continuing till the completion of the whirling.

b. It has been stated [§ 111,] that there are vegetable Bodies. He repels the objection of the atheist, that, in the case in.

Aph. 121.* Knowledge of the external is not indispensable [to constitute a Body]: trees, shrubs, climbers, annuals, trees with invisible flowers, grasses, creepers, &c., [which have internal consciousness], are, also, sites of experiencer and experience; as in the former case.

a. There is no necessity that that only should be a Body, in which there is knowledge of the external; but it is to be held that the being a Body, in the form of being the site of experiencer and experience, belongs also to trees, &c., which have internal consciousness; because, 'as in the former case,' meaning the putrescence already mentioned [see § 114], of the Bodies of men, &c., [which takes place] in the absence of the superintendence of an experiencer [the living soul], even in the same way do withering, &c., take place in the Bodies of trees, &c., also: such is the meaning. And to this effect there is Scripture.

Aph. 122.* And from the Legal Institutes [the same fact may be inferred, *viz.*, that vegetables have bodies and are conscious].

a. But then, from the fact that trees, &c., also, are thus conscious, we should find merit and demerit accruing to them. To this he replies:

Aph. 123.* Not merely through a Body is there susceptibility of Merit and Demerit; for Scripture tells us the distinction.

a. The vital spirit is not liable to the production of Merit and Demerit through a *Body* merely. Why? 'For Scripture tells us the distinction:' because we are told, in Scripture, that the liability results just from the being distinguished by a Brahmanical Body, or the like [animal body, not vegetable]. Such is the meaning.

b. Showing that the liability to Merit and Demerit is solely through the *kind* of Body, he mentions how Body is of three kinds:

Aph. 124.* Among the three there is a threefold distribution; the Body of merit, the Body of experience, and the Body of both.

a. There is a threefold distribution of Body 'among the three,' *i.e.*, among those highest, lowest, and intermediate,—all living beings,—*viz.*, the Body of merit, the Body of experience, and the Body of both: such is the meaning. Of these, a Body of merit belongs to the preeminent sages; a Body of experience, to Indra and others, and to things immovable, &c.; and a Body of both, to the royal sages. Here the division is [not exhaustive, but] into three, because of the preeminence [of these]; for, otherwise, we should have all alike possessed of a Body of experience, [like Indra].

b. He mentions also a fourth Body:

Aph. 125.* Not any one [of these], moreover, is that of the apathetic.

a. That is to say: the Body which belongs to the ascetics is different from all these three; such as was that of Dattatreya, Ja

abharata, and others; for they possessed bodies consisting of mere knowledge.

b. In order to establish the non-existence of a Lord, which was stated before, he disproves the eternity of knowledge, desire, action, &c., which is accepted by others [as existing in the case of the Lord]:

Aph. 126.* Eternity does not [as is alleged by those who wish to establish the existence of a Lord,] belong to knowledge, &c., even in the case of the particular site, [*viz.*, that of the supposed Lord]; as is the case with fire.

a. That is to say: just as we infer, from the example of ordinary fire, that the empyrean fire, also, is not eternal.

Aph. 127.* And, because the site [*viz*. the supposed Lord] is unreal, [it matters not, in the present instance, whether knowledge, &c.. may be eternal, or not].

a. But then, in that case, how can it, indeed, be possible that there should arise Omniscience, &c., adequate to the creation of the universe; since we do not behold, in mundane life, *such* superhuman powers [though we do see some,] arising from penance and the rest [of the alleged means of acquiring superhuman powers]? To this he replies:

Aph. 128.* The superhuman powers of concentration, just like the effects of drugs, &c., are not to be gainsaid.

a. That is to say: by the example of the effects of drugs, &c., even the superhuman powers of assuming atomic magnitude, &c., which result from concentration, and are adapted to the work of creation, &c., are established.

b. He refutes him who asserts that Thought belongs to the Elements; since this is hostile to the establishment [of the existence] of Soul:

Aph. 129.* Thought does not belong to the Elements; for it is not found in them separately, or, moreover, in the state of combination,—or, moreover, in the state of combination.

a. That is to say: Thought does not exist in the five Elements, even when in the state of combination; because we do not find Thought in them, generally, at the time of disjunction; [and there can be nothing in the product which does not preexist in the cause].

BOOK VI

Having explained, in four Books, all the matter of the Institute, and having, in the Fifth Book, thoroughly established it, by refuting the opinions of opponents, now, in a Sixth Book, he recapitulates the same matter, which is the essence of the Institute, while condensing it. For, in addition [to what has preceded], an enumeration of the matters before mentioned, namely, a summary, having been composed, learners acquire an undoubting, accurate, and more solid knowledge; so that, therefore, reiteration is not here to be imputed as a fault; because the method is that of fixing a stake, [*viz.*, by repeated blows], and because arguments, &c., not previously stated, are adduced.

Aph. 1.* Soul is; for there is no proof that it is not.

a. Soul really is existent, generically; since we are aware of this, that 'I think;' because there is no evidence to defeat this. Therefore, all that is to be done is to *discriminate* it [from things in general]. Such is the meaning.

b. The discrimination of it he establishes by means of two proofs:

Aph. 2.* This [Soul,] is different from the Body, &c.; because of heterogeneousness, [or complete difference between the two].

Aph. 3.* Also because it [Soul,] is expressed by means of the sixth [or possessive,] case.

a. That is to say: Soul is different from Body, &c., also because the learned express it by the possessive case, in such examples as, 'This [is] my body,' 'This [is] my understanding;' for the possessive case would be unaccountable, if there were absolute non-difference [between the Body, or the like, and the Soul, to which it is thus attributed as a possession].

b. But then, suppose that this, also, is like the expressions, 'The Soul's Thought' [Soul and Thought being identical], 'Rahu's head' [the trunkless Rahu being *all* head], 'The statue's body,' &c. To this he replies:

Aph. 4.* It is not as in the case of the statue; because there is [there] a contradiction to the evidence which acquaints us with the thing.

a. This expression by means of the possessive case, [*viz.*, 'My body' (§ 3)] is not like 'The statue's body,' &c. In such a case as 'The statue's body,' there is a mere fiction; 'for it is contradicted by the evidence which acquaints is with the thing;' [sense being the evidence that there is here no body other than the statue]. But, in such an expression as 'My body,' there is no contradiction by evidence; for the contradiction, by Scripture and other evidences, is only in supposing the Body to be the Soul. Such is the meaning.

b. Having settled that Soul is different from Body, &c., he settles its emancipation:

Aph. 5.* Through the entire cessation of pain there is done what was to be done.

a. But then, since there is an equality of gain and loss, inasmuch as, through the cessation of Pain there is the ceasing of Pleasure, also, *that* cannot be Soul's aim. To this he replies:

Aph. 6.* Not such desire for pleasure is there to Soul, as there is annoyance from Pain.

a. And so the aversion to Pain, having excluded also the desire for Pleasure, gives rise to a wish for he cessation of Pain simply; so that there is not an equality of gain and loss: [but a clear gain, in the desired release].

b. He declares that Soul's aim is simply the cessation of Pain; because Pain is, indeed, abundant, in comparison of Pleasure:

Aph. 7.* For [only] some one, somewhere, is happy.

a. Among innumerable grasses, trees, brutes, birds, men, &c., very few,—a man, a god, or the like,—are happy: such is the meaning.

Aph. 8.* It [Pleasure,] is also mixed with Pain, therefore the discriminating throw it to the side of [and reckon it as so much,] Pain.

a. He rejects the opinion that Soul's aim is not the simple cessation of Pain, but this [cessation] tinctured with Pleasure:

Aph. 9.* If you say that this [cessation of Pain] is not Soul's aim, inasmuch as there is no acquisition of Pleasure, then it is not as you say; for there are two kinds [of things desired].

a. For we see, amongst men, quite a distinct aspiration: [the first,] 'May I be happy;' [the second,] 'May I not be miserable;' [and the latter is *our* conception of beatitude].

b. He ponders a doubt:

Aph. 10.* The Soul [some one may suggest,] has *no* quality; for there is Scripture for its being unaccompanied, &c.

a. Therefore the cessation of Pain, indeed, [a property which does not belong to it,] cannot be Soul's aim: such is the meaning.

b. He clears up this [doubt]:

Aph. 11.* Though it [the Pain,] be the property of something else, yet it exists in it [the Soul,] through non-discrimination.

a. Though qualities, *viz.*, pleasure, pain, &c., belong [only] to the *Mind*, they exist, *i.e.*, they abide, in the shape of a *reflexion*, in it, *viz.*, in Soul, 'through non-discrimination,' as the cause, owing to the conjunction of Nature with Soul: such is the meaning. And this has been set forth in the First Book.

b. The binding of Soul by the qualities [or fetters,] arises from non-discrimination: but from what does non-discrimination arise? With reference to this, he says:

Aph. 12.* Non-discrimination [of Soul from Nature] is beginningless; because, otherwise, two objections would present themselves.

a. For, had it a beginning, then, if [first,] it arose quite spontaneously, bondage might befall even the liberated; and, if [secondly,] it were produced by Desert, &c., there would be a *regressus in infinitum*, inasmuch as we should have to search for another [previous instance of] non-discrimination, to stand as the cause of [that] Desert, &c., also: such is the meaning.

b. And then, if it be without beginning, it must be everlasting. To this he replies:

Aph. 13.* It [non-discrimination, cannot be everlasting [in the same manner] as the soul is; else, it could not be cut short, [as we affirm that it can be].

a. It is not everlasting, indivisible, and beginningless, in the same way as the soul is; but it is beginningless, in the shape of an on-flow [which may be stopped]. For, otherwise, the cutting short of a beginningless entity would, as is established by Scripture, be unfeasible, [though the beginningless antecedent non-entity of a given jar may be readily understood to terminate, on the production of the jar]. Such is the meaning.

b. Having stated the cause of [Soul's] Bondage, he states the cause of Liberation:

Aph. 14.* It [Bondage,] is annihilable by the alloted cause, [*viz*., discrimination of Soul from Nature]; as darkness is [annihilable by the allotted cause, *viz*., Light].

Aph. 15.* Here, also, [*viz*., in the case of Bondage and Discrimination, as in the case of Darkness and Light,] there is allotment, [as is proved] both by positive and negative consociation; [Liberation taking place where Discrimination is, and not where it is not].

a. He reminds [us] of what was mentioned in the first Book, *viz*., that Bondage cannot be innate, &c:

Aph. 16.* Since it cannot be [accounted for] in any other way, it is non-discrimination alone that is [the *cause* of] Bondage, [which cannot be innate].

a. 'Bondage' here means the *cause* of Bondage, named the conjunction of pain. The rest is plain.

b. But then, since liberation, also, from its being a product, is liable to destruction, Bondage should take place over again. To this he replies:

Aph. 17.* Further, Bondage does not again attach to the liberated; because there is Scripture for its non-recurrence.

Aph. 18.* Else, it [liberation,] would not be the Soul's aim, [which it is].

a. He states the reason why this is not Soul's aim:

Aph. 19.* What happened to both would be alike, [if liberation were perishable].

a. That is to say: there would be no difference between the two, the liberated and the bound; because of their being alike liable to future bondage; and, therefore, such [perishable emancipation] is not Soul's aim, [but emancipation final and complete].

b. But then, in that case, if you acknowledge that there is a *distinction* between the bond and the free, how is it that you have asserted [Book I., § 19,] the *eternal* freedom [of *all* souls alike]? To this he replies:

Aph. 20.* Liberation is nothing other than the removal of the obstacle [to the Soul's recognition of itself as free].

a. But then, in that case, since Bondage and Liberation are *unreal*, Liberation must be contradictory to the texts, &c., which set forth what is Soul's aim, [as some positive and real acquisition, not merely the removal of a screen]; to which he replies:

Aph. 21.* Even in that case, there is no contradiction.

a. That is to say: 'even in that case,' *i.e.*, even if Liberation consists [only] in the removal of an obstacle, there is no contradiction in its being Soul's aim.

b. But then, if Liberation be merely the removal of an obstacle, then it should be accomplished through mere *hearing* [of the error which stands in the way]; just as a piece of gold on one's neck, [which one has sought for in vain, while it was] withheld from one by ignorance [of the fact that it has been tied round one's neck with a string], is attained, [on one's hearing where it is]. To this he replies:

Aph. 22.* This [attainment of Liberation, on the mere *hearing* of the truth,] is no necessity; for there are three sorts of those competent [to apprehend the truth; but not all are qualified to appropriate it, on merely hearing it].

a. He mentions that not mere *hearing* alone is seen to be the cause of knowledge, but that there are others, also:

Aph. 23.* Of others [*viz.*, other means besides hearing] for the sake of confirmation, [there is need].

a. He speaks of these same other means:

Aph. 24.* There is no [absolute] necessity that what is steady and promoting ease should be a [*particular*] posture, [such as any of those referred to in Book III., § 34].

a. That is to say: there is no necessity that a 'posture' should be the 'lotus-posture,' or the like; because whatever is steady and promotes ease is a [suitable] 'posture.'

b. He states the principal means [of Concentration]:

Aph. 25.* Mind without an object is Meditation.

a. That is to say: what Internal Organ is void of any modification, *that* is 'Meditation,' *i.e.*, Concentration, in the shape of exclusion of the modifications of Intellect: by reason of the identity [here,] of effect and cause, the word 'cause' is employed for 'effect.' For it will be declared how Meditation effects this [exclusion of the modifications of Intellect].

b. But then, since Soul is alike, whether there be Concentration or Non-concentration, what have we to do with concentration? Having pondered this doubt, he clears it up:

Aph. 26.* If you say that even both ways there is no difference, it is not so: there is a difference, through the excluion [in the one case,] of the tinge [of reflected pain which exists in the other case].

a. But how can there exist a *tinge* in that which is unassociated [with anything whatever, as Soul is alleged to be]? To this he replies:

Aph. 27.* Though it [Soul,] be unassociated, still there is a tingeing [reflexionally,] through Non-discrimination.

a. That is to say: though there is not a *real* tinge in that which is unassociated [with tincture, or anything else], still there is, *as it were*, a tinge; hence the tinge is treated as simply a *reflexion*, by those who discriminate the tinge [from the Soul, which it delusively seems to belong to].

b. He explains this same:

Aph. 28.* As is the case with the Hibiscus and the crystal [Book I., § 19, *c*.], there is not a tinge, but a fancy [that there is such].

a. He states the means of excluding the aforesaid tinge:

Aph. 29.* It [*viz.*,the aforesaid tinge,] is debarred by Meditation, Restraint, Practice, Apathy, &c.

a. He shows the means settled by the ancient teachers, in regard to the exclusion—through Meditation, &c., lodged in the Mind,—of the tingeing of Soul:

Aph. 30.* It is by the exclusion of dissolution and distraction, say the teachers.

a. That is to say: through the removal, by means of meditation, &c., of the Mind's condition of [being dissolved in] Sleep, and condition of [waking] Certainty, &c., there takes place also the exclusion of the tingeing of Soul by the condition; because, on the exclusion of any [real] object, there is the exclusion also of its reflexion: so say the ancient teachers.

b. He states that there is no compulsion that Meditation, &c., should take place in caves and such places:

Aph. 31.* There is no rule about localities; for it is from tranquillity of Mind.

a. That is to say: Meditation, or the like, results simply 'from tranquillity of Mind.' Therefore, such a place as a cave is not indispensable for it.

b. The discussion of Liberation is completed. Now, with an eye to the unchangeableness of Soul, he handles compendiously the cause of the world:

Aph. 32.* Nature is the primal material; for there is Scripture [to the effect] that the others are products.

a. That is to say: since we learn, from Scripture, that Mind, &c., are products, Nature is established under the character of the radical cause of these.

b. But then, let *Soul* be the material. To this he replies:

Aph. 33.* Not to Soul does this [*viz.*, the material of the world,] belong, though it be eternal; because of its want of suitableness.

a. That is to say: suitableness to act as material implies possession of qualities, and the being associable: [and,] by reason of the absence of both of these, Soul, though eternal, [and, therefore, no product,] cannot serve as material.

b. But then, since, from such Scriptural texts as, 'Many creatures have been produced from Soul,' we may gather the fact that Soul is a cause, the assertions of an illusory creation, &c., ought to be accepted. Having pondered this adverse suggestion, he replies:

Aph. 34.* The despicable sophist does not gain [a correct apprehension of] Soul; because of the contradictoriness [of his notions] to Scripture.

a. That is to say: the various views, in regard to Soul's being a cause, which are conceivable are, all, opposed to Scripture; therefore, the lowest of the bad reasoners, and others, who are accepters thereof, have no knowledge of the nature of Soul. Hence it is to be understood that those, also, [*e.g.*, the *Naiyayikas*,] who assert that Soul is the substance of the qualities Pleasure, Pain, &c., are quite illogical; these, also, have no correct knowledge of Soul. And, if it be asserted that Soul is a cause [of the world], just as the sky is the recipient cause of the clouds, &c., [and stands, towards it, in the relation of a cause, in so far as, without the room afforded by it, these could not exist], then we do not object to *that*; for, what we deny is only that there is transformation [of soul, as material, into the world, as product].

b. Since we see, that, in the case of things motionless, locomotive, &c., the material cause is nothing else than *earth*, &c., how can *Nature* be the material of all? To this he replies:

Aph. 35.* Though but mediately [the cause of products], Nature is inferred [as the ultimate cause of the intermediate causes,]; just as are Atoms, [by the *Vaisheshikas*].

Aph. 36.* It [Nature,] is all-pervading; because [its] products are seen everywhere.

a. But then, only if it be *limited*, can it be said that, 'Wherever a product arises, there does it [Nature,] go [or act];' [for what is unlimited] and fills all space, can find no other space to *move* into]. To this he replies:

Aph. 37.* Though motion may attach to it, this does not destroy its character as ultimate cause; just as is the case with Atoms.

a. 'Motion' means action. Though it be present, this does not prevent its [Nature's,] being the radical cause; just as is the case with the earthy and other Atoms, according to the opinion of the *Vaisheshikas*: such is the meaning.

Aph. 38.* Nature is something in addition to the notorious [nine Substances of the *Naiyayikas*]: it is no matter of necessity [that there should be precisely nine].

a. And the argument, here, is the Scriptural declaration, that *eight* [of the pretended primitive substánces] are *products*: such is the import.

Aph. 39.* Purity and the others are not *properties* of it [*viz*., Nature]; because they are its *essence*.

a. That is to say: Purity and the other Qualities are not *properties* of Nature; because they are what *constitutes* Nature.

b. He determines the motive of Nature's energizing; since, if we held the energizing to be without a motive, Emancipation would be inexplicable:

Aph. 40.* Nature, though it does not enjoy [the results of its own energizing], creates for the sake of Soul; like a cart's carrying saffron, [for the use of its master. See Book III., § 58].

a. He states the concomitant cause of diversified creation:

Aph. 41.* The diversity of creation is in consequence of the diversity of Desert.

a. But then, granting that *creation* is due to Nature, yet whence is *destruction*? For a couple of opposite results cannot belong to one and the same cause. To this he replies:

Aph. 42.* The two results are through equipoise and the reverse of equipoise.

a. Nature is the triad of Qualities, *viz*., Purity, &c.; and their 'reverse of equipoise' is their aggregation in excess or defect; the absence of this [reverse of equipoise] is 'equipoise:' through these two causes two opposite results, in the shape of creation and destruction, arise from one and the same: such is the meaning.

b. But then, since it is Nature's attribute to create, there should be the mundane state, even after [the discriminative] knowledge, [which, it is alleged, puts an end to it]. To this he replies:

Aph. 43.* Since [or when,] the emancipated has understood [that he never was really otherwise], Nature does not create;

just as, in the world, [a minister does not toil,when the king's purpose has been accomplished].

a. But then, Nature does not rest from creating; for we see the mundane condition of the ignorant: and so, since Nature goes on creating, to the emancipated, also, Bondage may come again. To this he replies:

Aph. 44.* Even though it [Nature,] may invade others [with its creative influences], the *emancipated* does not experience, in consequence of the absence of a concurrent cause, [*e.g.*, Non-discrimination, in the absence of which there is no reason why the emancipated should be subjected to Nature's invasion].

a. But then, *this* arrangement could be possible then, [only] if there were a *multiplicity* of souls: but *that* is quite excluded by the text of the non-duality of Soul. Having pondered this doubt, he says:

Aph. 45.* The multeity of Soul [is proved] by the distribution [announced by the *Veda* itself].

a. That is to say: the multeity of Soul is proved, absolutely, by the distribution of Bondage and Emancipation mentioned in such Scriptural texts as, 'Whoso understand this, these are immortal, while others experience only sorrow.'

b. But then, the distribution of Bondage and Liberation may be through the difference of adjunct. To this he replies:

Aph. 46.* If [you acknowledge] an adjunct [of Soul], then, on *its* being established, there is *duality*, [upsetting the dogma founded on in § 44].

a. But then, the adjuncts, moreover, consist of 'Ignorance,' [which, according to the Vedanta, is no reality]; so that by these there is no detriment to [the Vedantic dogma of] non-duality. With reference to this doubt, he says:

Aph. 47.* Even by the two the authority is contradicted.

a. Thst is to say: even by acknowledging the two, *viz.*, Soul and Ignorance, a contradiction is constituted to the text, [which is alleged as] the authority for non-duality.

b. He states another couple of objections, also:

Aph. 48.* The prima facie view [of the Vedanta] is not [to be allowed any force, as an objection]; because, by [admitting] two, [*viz.*, Soul and Ignorance], there is no opposition [to our own dualistic theory of Soul and Nature]: and the subsequent [dogma, *viz.*, that one single Soul is the only reality, is not to be allowed]; because of the non-existence of a proof, [which, if it *did* exist, would, along with Soul, constitute a duality].

a. But then, Soul will be demonstrated by its self-manifestation. To this he replies:

Aph. 49.* [And,] in its [Soul's,] being demonstrated by the light [of itself as you Vedantis say it is], there is the [unreconciled] opposition of patient and agent [in one, which is a contradiction].

a. That is to say: if Soul be demonstrated by the light which Soul consists of, there is the 'opposition of patient and agent' [in one].

b. But then, there is no contradiction [here,] between patient and agent; because it [the Soul], through the property of light which is lodged in it, can, itself, furnish the relation to itself; just as the *Vaisheshikas* declare, that, through the intelligence lodged in it, it is, itself, *object* to itself. To this he replies:

Aph. 50.* This [Soul], in the shape of Thought, discrepant from the non-intelligent, reveals the non-intelligent.

a. But then, in that case, if duality be established in accordance with proofs, &c., what becomes of the Scriptural text declaring non-duality? To this he replies:

Aph. 51.* There is no contradiction to Scripture [in our view]; because that [text of Scripture which seems to assert absolute non-duality] is [intended] to produce apathy in those who have desires, [and who would be better for believing in 'the *nothingness* of the things of time'].

a. He tells us that the assertors of non-duality are to be shunned, not only for the reason above mentioned, but, also, because of the non-existence of evidence to convince us that the world is unreal:

Aph. 52.* The world is real; because it results from an

unobjectionable cause, and because there is [in Scripture,] no debarrer [of this view of the matter].

a. We see, in the world, that no reality belongs to dream-objects, or to the [fancied] yellowness of [invariably white] conch-shells, and the like; inasmuch as these are results of the internal organ, &c., when [not normal, but] injured by [*i.e.*, under the injurious influence of] Sleep, &c.: and this is not [the state of things] in the [waking] Universe, in which Mind is the first, [according to Book I., § 71].

b. He declares that the Universe is real, not merely in its existent state [at any given instant], but, also, always:

Aph. 53.* Since it cannot be [accounted for] in any other way, manifestation [of whatever is manifested] is of what is real, [*i.e.*, of what previously existed].

a. That is to say: since, through the aforesaid reasons, it is impossible that the unreal should come into existence, what does come into existence, or is manifested, is what really existed [previously,] in a subtile form.

b. Though [it is declared that] the being the agent and the being the experiencer belong to diverse subjects, be asserts the distribution [of agency to Self-consciousness, and of experience to Soul,] by two aphorisms:

Aph. 54.* Self-consciousness, not Soul, is the agent.

Aph. 55.* Experience ceases at [discrimination of] Soul, [as being quite distinct from Nature]; since it arises from its [Soul's,] Desert, [which is not, *really*, Soul's, but which, while Non-discrimination lasts, is made over to Soul; just as the fruits of the acts of a king's ministers are made over to the king].

a. He shows the reason for what was stated before, *viz*., that cessation of action does not result from enterings into the world of Brahma:

Aph. 56.* Even in the world of the moon, &c., there is return [to munane existence]; because of there really being a cause [of such return].

a. 'A cause,' *viz*., Non-discrimination, Desert, &c.

b. But then, through the counsels of the persons dwelling in these various [supermundane] worlds, there ought to be no return [to mundane existence]. To this he replies:

Aph. 57.* Not by the counsel of [supermundane] people is there effectuation [of Emancipation]; just as in the former case, [the case, *viz*., of counsel given by mundane instructors].

a. But, in that case, what becomes of the text that there no return from the world of Brahma? To this he replies:

Aph. 58.* There is Scripture [declaratory] of Emancipation, [on going to the world of Brahma]; this [Emancipation] being effected [more readily in that world than in this, but only] by intermediacy [of the appropriate means].

a. He alleges the Scriptural text of Soul's *going* [to the locality where it is to experience], even though it be all-filling, [and can, therefore, have no place into which to move]:

Aph. 59.* And, in accordance with the text of its 'going,' though it [Soul,] is all-pervading, yet, in time, it reaches its place of experience [or body], through conjunction with an adjunct; as in the case of Space.

a. For, as Space, though it is all-pervading, is spoken of as moving to some particular place, in consequence of its conjunction with an adjunct, such as a jar, [when we say 'the space occupied by the jar is moved to the place to which the jar is carried'], just so is it [here].

b. He expounds the statement, that the site of experience [the body,] is formed through the superintendence of the experiencer, [Soul]:

Aph. 60.* This [constitution of a body] is not accomplished in the case of what is [organic matter] not superintended [by Soul]; because we find putrefaction [in organic matter where Soul is absent].

a. But then, let the construction of a site of experience [or a body,] for Experiencers [*i.e.*, Souls,] take place without any superintendence at all, through Desert. To this he replies:

Aph. 61.* If you say that [independently of any superintendence,] it is through Desert [that a Body is formed, it is not so]; since what is unconnected [with the matter to be operated upon] is incompetent thereto; as is the case with [unapplied] water, &c., in respect of a plant.

a. That is to say: because it is impossible that Desert, which is not directly conjoined with the semen and other [elements of the Body], should operate *through Soul*, in the construction of the Body, &c.; just as it is for water, &c., *unconnected* with the seed, to operate *through tillage*, &c., in the production of a plant.

b. According to the system of the *Vaisheshikas* and others, it is settled that Soul is the superintendent, [in the construction of the Body], *in virtue of its being conjoined with Desert*. But he tells us, that, in his own doctrine, since Desert, &c., are not properties of Soul, the Soul cannot, *through these*, be the cause [of the Body]:

Aph. 62.* For this is impossible [*viz.*, that the Soul should, *through its Desert*, &c., be the cause of Body]; because it has *no* qualities for these [*viz.*, Desert, &c.,] are properties of Self-consciousness, [not of Soul].

a. And so, in *our* opinion, it is settled that Soul superintends [in the causing of the Body,] quite directly, by conjunction simply, without reference to anything intermediate: such is the import.

b. But, if Soul be all-pervading, then the limitedness of the living soul, which is set forth in Scripture, is unfounded. To repel this doubt, he says:

Aph. 63.* The nature of a living soul belongs to that which is qualified, [not to soul devoid of qualities, as is proved] by direct and indirect argument.

a. To be a living soul is the being possessed of the vital airs; and this is the character of the soul distinguished by peraonality, not of pure Soul, [which is unlimited].

b. Desiring, now, to set forth the difference between the products of Mind [or the Great Principle,] and of Self-consciousness, he first states the products of Self-consciousness:

Aph. 64.* The effectuation of works is dependent on the

agent Self-consciousness, not dependent on a Lord, [such as is feigned by the *Vaisheshikas*]; because there is no proof [of the reality of such].

a. By this aphorism are set forth, as are also established by Scripture and the Legal Institutes, the creative and the destructive agencies of Brahma and Rudra [respectively], owing to their adjunct, Self-consciousness, [or peronality].

b. But then, grant that Self-consciousness is the maker of the others, still who is the maker of Self-consciousness? to this he replies:

Aph. 65.* It is the same as in the arising of Desert.

a. Just as, at the creations, &c., the manifestation of Desert, which sets Nature energizing, results solely from the particular *time*,—since, if we were to suppose other Desert as the instigator of this, we should have an infinite regress,—just so Self-consciousness arises from *time* alone. as the cause; but there is not another maker thereof, also: thus, the two [cases] are alike: such is the meaning.

Aph. 66.* The rest is from Mind, [the Great Principle].

a. What is other than the products of Self-consciousness [or personality], *viz*., Creation, &c., that, *viz*., Preservation, &c., results from the Great Principle alone; because, inasmuch as it consists of pure Goodness, having no Conceit, Passion, &c., it is moved solely by benevolence towards others: such is the meaning. And by this aphorism is established the character, as Preserver, of Vishnu, owing to the Great Principle, as adjunct [of the soul, which, without adjunct, would neither create, preserve, nor destroy (see § 64)].

b. It has been stated, before, that the relation of Nature and Soul, as experienced and experiencer, is caused by Non-discrimination [of the one from the other]. Here, what is Non-discrimination, itself, caused by? With reference to this doubt, he states that all philosophers reject, in common, the doubt whether we should have an infinite regress, on the supposition of a *stream* of Non-discrimination; because *this* [regress] is *valid*; [since an infinite regress which is in conformity with the truth is no sound cause of objection]:

Aph. 67.* The relation of possession and possessor, also, if attributed [as it is by some,] to Desert, in the case of Nature [and Soul], like [the relation of] seed and plant, [which takes the shape of an infinite regress of alternants], is beginningless.

Aph. 68.* Or [the case is, likewise, one of infinite regress,] if it [the relation between Nature and Soul,] be attributed to Non-discrimination [of Soul from Nature], as Panchashikha [holds].

Aph. 69.* [The case is the same,] if, as the teacher Sanandana does, we attribute it [the relation between Nature and Soul,] to the Subtile Body, [which, in the shape of its elemental causes, attends Soul, even during the periodical annihilations of the world].

a. He sums up the import of the declarations of the Institute:

Aph. 70.* Be that the one way, or the other, the cutting short thereof [*viz*., of the relation between Nature and Soul,] is Soul's aim; the cutting short thereof is Soul's aim.

Bibliography

Ahmad, Imtiaz: *State and Foreign Policy: India's Role in South Asia*, Vikas, New Delhi, 1993.

Ainslee T. Embree: *Imagining India: Essays on Indian History*, Oxford Univ. Press, Delhi, 1989.

Ames, Roger T.: *The Art of Rulership: A Study in Ancient Chinese Political Thought*, University of Hawaii Press, Honolulu, 1983.

Apte, V. M.: *Social and Religious Life in the Grihya Sutras*, Rawat Press, Bombay, 1954.

Archer, W. G.: *Kangra Painting*, Faber and Faber, London, 1956.

Artola, George T.: *The Banner of Kamadeva and other Topics of Sanskrit Literature and Indian Culture*, Popular Prakashan, Bombay, 1977.

Austin, Granville: *The Indian Constitution: Cornerstone of a Nation*, Clarendon Press, Oxford, 1966.

Baker, Christopher J.: *An Indian Rural Economy: The Tamiland Countryside*, Clarendon Press, Oxford, 1984.

Beach, Milo Cleveland: *The New Cambridge History of India, I.3: Mughal and Rajput Painting*, Cambridge University Press, Cambridge, 1992.

Bhardwaj, Surinder Mohan: *Hindu Places of Pilgrimage in India: A Study in Cultural Geography*, University of California Press, Berkeley, 1973.

Bhattacharjee, Arun: *Rajiv Gandhi: Life and Message*, Ashish, New Delhi, 1992.

Bowes, Pratima: *Hindu Intellectual Tradition*, Allied Publishers, New Delhi, 1977.

Bromley, David and Larry Shinn: *Krishna Consciousness in the West,* Bucknell University Press, Lewisburg, 1989.

Brown, Judith M.: *Modern India: The Origins of an Asian Democracy,* Oxford University Press, New York, 1994.

Cauquelin, J.: *Asian Values. Encounters with Diversity*, Surrey, Richmond, 2000.

Collect, S. D.: *The Life and Letter of Raju Rahmmohun Roy*, Kali Ram Press, Calcutta, 1962.

Cunningham, Joseph Davey: *History of the Sikhs, From the Origins of the Nation to the Battles of the Sutlej*, Sultan Chand, Delhi, 1955.

Das, M.N.: *India under Morley and Minto: Politics Behind Revolution, Repression, and Reforms*, Allen and Unwin, London, 1964.

De, Sushil Kumar: *Ancient Indian Erotics and Erotic Literature*, Firma K. L. Mukhopadhyay, Calcutta, 1959.

Dehejia, Vidya: *Antal and Her Path of Love: Poems of a Woman Saint from South India*, State University of New York Press, Albany, 1990.

Diwan, Paras: *Human Rights and the Law: Universal and Indian,* Deep and Deep, New Delhi, 1996.

Dixit, Prabla: *Communalism: A Struggle for Power*, Orient Longman, New Delhi, 1981.

Eno, Robert: *The Confucian Creation of Heaven: Philosophy and the Defense of Ritual Mastery*, SUNY Press, Albany, 1990.

Erickson, Erik H.: *Gandhi's Truth: On the Origins of Militant Nonviolence*, Norton, New York, 1970.

Farquhar, J. N.: *A Primer of Hinduism,* Faber & Faber, London, 1912.

Farrand, Livingston: *Traditions of the Quinalut Indians,* American Museum of Natural History, New York, 1902.

Gambhirananda, Swami: *Brahma Sutra Shamkar Bhasya,* Adavita Ashrama, Calcutta, 1977.

Heinsath, Charles: *Indian Nationalism and Hindu Social Reform*, Princeton University Press, Princeton, 1964.

Howard Schwartz: *Reimaging the Bible: The Storytelling of the Rabbis,* Oxford University Press, New York, 1998.

John Drew: *India and the Romantic Imagination,* Oxford Univ. Press, London, 1987.

John, Elsner: *Pilgrimage: Past and Present in the World Religions,* Harvard University Press, Cambridge, 1995.

Kapur, Rajiv: *Sikh Separatism: The Politics of Faith*, Vikas, New Delhi, 1987.

Lal, Vinay: *Fingerprinting Popular Culture: The Mythic and the Iconic in Indian Cinema*, Oxford, Delhi, 2006.

Lamba, B.: *Fauna of Corbett National Park,* ZSI, Calcutta, 1987.

Louis, Fischer: *Essential Gandhi: An Anthology of His Writings,* Vintage, New York, 1983.

MacDonell, Arthur Anthony: *History of Sanskrit Literature*, South Asia Books, Columbia, 1990.

Mahalingam, T.V.: *Administration and Social Life under Vijayanagar*, University of Madras, Madras, 1969.

Michael A. Sells: *Early Islamic Mysticism: Sufi, Quran, Miraj, Poetic and Theological Writings*, Paulist Press, New York, 1996.

Misra, Satya Swarup: *The Aryan Problem: A Linguistic Approach*, Munshiram Manoharlal, New Delhi, 1992.

Morris M. Faierstein: *Jewish Mystical Autobiographies: Books of Visions and Book of Secrets*, Paulist Press, New York, 1999.

Narasimhan, C.V.: *The Mahabhárata: An English Version Based on Selected Verses*, Columbia University Press, New York, 1965.

Organ, Troy W.: *Hinduism: Its Historical Development,* Barron's Educational Series, New York, 1974.

Piatigorsky, Alexander: *The Buddhist Philosophy of Thought,* Rowman & Allenheld, London, 1984.

Potter, Karl: *Guide to Indian Philosophy*, G.K. Hall, Boston, 1988.

Prajnanananda, Swami: *Schools of Indian Philosophical Thought,* Firma K. L. Mukhopadhyay, Calcutta, 1973.

Ranade, R. D.: *Mysticism in Maharashtra*, MacMillan, Delhi, 1933.

Reuven, H.: *The Classic Midrash: Tannaitic Commentaries on the Bible*, Paulist Press, New York, 1995.

Rudolf Otto: *Mysticism East and West: A Comparative Analysis of the Nature of Mysticism*, MacMillan, New York, 1969.

Sankaia, H.D.: *Aspects of Indian History and Archeology*, B.R. Publishing, Delhi, 1977.

Sivaramamurti, C.: *Sri Lakshmi in Indian Art and Thought*, Kanak, New Delhi, 1982.

Thapar, Romila: *Ashoka and the Decline of the Mauryas*, Oxford University Press, London, 1961.

Thomas, P.: *Kama Kalpa or the Hindu Ritual of Love,* D. B. Taraporevala Sons, Bombay, 1960.

Thompson, Edward: *The Making of the Indian Princes*, Oxford University Press, London, 1943.

Tobias, Michael: *A Day in the Life of India,* Collins Publishers San Francisco, San Francisco, 1996.

Trimingham, J.: *Sufi Orders in Islam*, Oxford University Press, New York, 1998.

Walters, Donald: *The Path: Autobiography of a Western Yogi (Swami Kriyananda),* Ananda Publications, Nevada City, 1977.

Wells, Oliver: *Salish Weaving Primitive and Modern,* Oliver Wells, Sardis, 1970.

Zimmer, Heinrich. *The Art of Indian Asia*, Pantheon, New York, 1955.

Zion, Bokser: *Abraham Isaac Kook: The Lights of Penitence*, Paulist Press, New York, 1978.

Index

A

B

C

O

P

R

S

T

U

V

W

Y

□□□